TECHNOLOGY
AND THE
EDUCATIONAL
WORKPLACE

EIGHTEENTH ANNUAL YEARBOOK
OF THE AMERICAN EDUCATION FINANCE ASSOCIATION
1997

TECHNOLOGY AND THE EDUCATIONAL WORKPLACE

Understanding
Fiscal Impacts

Editor
Kathleen C. Westbrook

CORWIN PRESS, INC.
A Sage Publications Company
Thousand Oaks, California

For information:

Corwin Press, Inc.
A Sage Publications Company
2455 Teller Road
Thousand Oaks, California 91320
E-mail: order@corwinpress.com

SAGE Publications Ltd.
6 Bonhill Street
London EC2A 4PU
United Kingdom

SAGE Publications India Pvt. Ltd.
M-32 Market
Greater Kailash I
New Delhi 110 048 India

Library of Congress Cataloging-in-Publication Data

Main entry under title:

Technology and the educational workplace: Understanding fiscal impacts
/ [edited by Kathleen C. Westbrook].
    p. cm. — (Annual yearbook of the American Education Finance
Association; v. 18)
    Includes bibliographical references and index.
    ISBN 0-8039-6561-3 (cloth: acid-free paper)
    1. Educational technology—Finance. 2. Educational innovations—
Finance. 3. Educational technology—Social aspects. 4. Educational
technology—Planning. 5. Educational technology—Government policy.
6. Distance education. I. Westbrook, Kathleen C. II. Series.
LB1028.3 .T4253   1997
379.1'224'0973—ddc21   98-8974

This book is printed on acid-free paper.

98 99 00 01 02 03 10 9 8 7 6 5 4 3 2 1

Production Editor: Sherrise M. Purdum
Editorial Assistant: Denise Santoyo
Typesetter: Christina M. Hill
Cover Designer: Michelle Lee

Contents

Preface

It is a unique opportunity to be selected as editor of an AEFA yearbook. Only once in a great while do we have the opportunity to have a "stage" on which to display and write about the things that are most important to us. This volume is one of those rare opportunities. For quite a few years now, I have been involved in the struggle to bring technology into the lives of our profession—both in the university and K-12 sectors. I have heard all of the clichés many times over about "We don't do that here," or "We can't afford it," or "We don't have the time to learn that"; yet, like that venerable comic-strip character Charlie Brown, I keep pounding my head against that proverbial tree, and like that childhood sage, I reply to the question, "Why do you keep banging your head against the tree?" with "Because it feels so good when I stop!" Some might say this is the half-empty approach; I prefer to believe it is the tenacious one.

Tenacity—that is what the authors of this volume exemplify. They have met the challenge of technology head on and, in their own ways are the victors in the ongoing battles. The fiscal issues identified in these stories are glaring, for they remind us of yet how far we need to go professionally as we make the final turn into the next century. They remind us also of the unsung heroism that lurks within each of us. These are stories from the field—in some cases literally from battle-fields—on the courage, spirit, and dedication that these pioneers foster within themselves. They are stories of hope, as they represent the finest thinking and hoping within the professions of research, analysis, and policy. They are also stories of frustrations—at the waste, lack of vision, and irresponsibility of those with the power to move our educational systems around this ever-shrinking globe into a more harmonious whole and with greater forward momentum.

We open our story with international songs. Maureen W. McClure's chapter on the GINIE project brings home the dedication and commitment of individuals in the United States to seek a humanistic solution to isolation and devastation via technologic means.

Emily Vargas-Baron shows how state financing of higher education partnerships helps to increase trade and economic development.

Next, a chapter by Sreben Dizdar and Cecilia Wandiga takes us to war-torn Bosnia-Herzegovina to keep education on the front burner for the next generation of children in that war-ravaged country. The fiscal devastation, graft, and political corruption do not prevent the education of its children. For these individuals, commitment to the future is more important than the present. Fiscal and personal sacrifices are negligible when compared to a future of uneducated or undereducated leaders.

In the section on state and local issues, Karen Fullerton cautions us on the mythic crutches we use to keep ourselves—and our institutions—from moving forward at more than a snail's pace and to maintain our comfortable status quo.

Next, the section offers two unique perspectives—one on "re-visioning" our state and federal roles and one on how teacher unions can become a creative part of the process. Both Faith E. Crampton (NEA) and F. Howard Nelson (AFT) bring the unique perspective of the two major teacher bargaining units to bear on these issues. Where and how does technology and its fiscal availability become a road-block to access and equity for children? How can teachers and parents use this medium to keep in touch with, rather than becoming more isolated from, their children and their children's schooling?

In Section III the focus turns to those affecting training and professional development in K-12 teacher, administrator, and higher educational settings. Here, Barbara Y. LaCost, Alan T. Seagren, and Sheldon L. Stick tell us about unique programs designed to model new ways of thinking, knowing, and behaving. We also hear from Hank Bromley, Stephen L. Jacobson, Michael L. Waugh, and Marianne Handler on how institutional resources can add to—or frustrate—the individuals trying to implement these programs, and how the clients—the teachers and faculty exposed to them—react to new service delivery systems and methodologies and where and how resources are needed in this process of change.

The last two chapters take us into the policy arena. Chapter 10 helps us see this entire picture from the state legislative point of view. Here, James F. Angevine shoots straight from the hip at those of us who work with legislators. He challenges us to engage in a little crystal ball gazing to envision and begin the dialogues on what the future could be—and what roles we can play before the final curtain.

The final chapter in this volume takes us from where we've been to where we're going. It challenges us as fiscal analysts to be drumbeaters, cajolers, and advocates of new ways of viewing the traditional area of fiscal resource allocation. Knowing what we know about the startup costs, staff development needs, and rates of obsolescence of technology requires us to be much more proactive and creative in how we approach the resource provisions, equity, and access issues that for so long have consumed our analysis. This volume was a great challenge—and opportunity. Without the unceasing cooperation of Corwin Press, Ann McMartin and the Corwin editorial staff, it would never have come to print. An incredible debt of gratitude is also owed to Dr. Mary McKeown-Moak, whose belief in this as a timely topic for an AEFA yearbook can never adequately be repaid. And last, to the two most wonderful professional colleagues and mentors anyone could ever hope to find—Dr. Maureen McClure and Dr. Gene McLoone—I am deeply indebted for their unceasing and unerring dedication to the quality of this volume and to their personal commitment and support during the hills and valleys that accompany any task of this magnitude.

Kathleen C. Westbrook

Introduction

On my desk is a wondrous photograph. I recently acquired it while working with the education specialist from the Adler Planetarium and Astronomy Museum. We were developing their new World Wide Web site for an Illinois State Board of Education Science Teacher Enhancement grant titled "Near and Far Sciences for Illinois." In this project, science teachers from around the state of Illinois will engage over the next 9 months in activities to develop their knowledge and teaching skills in the areas of astronomy, geology, and meteorology. The Web site (http://newton.dep.anl.gov/nfsci/index.htm) will be used in conjunction with their regional activities and in collaboration with the scientists who will be working on the project. The photograph is amazing in its color, clarity, and definition. It is a photograph of the Eagle Nebula. This nebula contains hydrogen and microscopic dust particles, which are the raw materials for new stars. It gets its name from the bird of prey with outstretched wings and talons. The wonderment of this photograph is that it exists at all. The Eagle Nebula is 2 million years old and located in the southern Milky Way—more than 7,000 lightyears from earth. Seven thousand lightyears—the mind cannot even comprehend such a distance. Yet, here I am looking at a photograph of stars being born at a distance I could not travel during my lifetime. Our own galaxy would fit comfortably inside one of the "eggs" (evaporating gaseous globules) that are the genesis of newly forming stars. The nebula itself has one branch that is nearly one lightyear in length from top to bottom (a lightyear is the distance light travels in a year, or 6 x 10^{12}, or 5 trillion 900 billion miles). How is this photograph possible? The answer: *technology.* The Hubble Space Telescope, for all its problems, has enabled researchers at Arizona State University to photograph this wondrous sight without ever leaving the comfort of their offices, cities, communities, or planet.

This is the ushering in of a "New Age." Society's advancement is no longer constrained by cosmological, geographical, or political restrictions. The basis of our lives, predicated on capital exchange and formation, is changing as are the sands of the desert, beneath our very feet. Not only computers, but technology itself is changing the fabric and meaning of culture, time, distance, and space. With a cellular phone no larger than a wallet, we can call halfway around the globe and place an order, select a product, or run a multinational corporation. In the wink of an eye, we can communicate with astronauts revolving about the earth in a space capsule to conduct experiments and repairs on satellites and ships in the void of outer space. Yet, with all this "instantaneous" communication, we find a global environment rife with turmoil this technology was supposed to ameliorate— war, famine, homelessness, poverty, urban and educational blight, racism, political and religious intolerance, and economic distress.

This 1997 Yearbook of the American Education Finance Association is unique. It is not simply another year's look at the ways we view education finance allocation formulas or develop distributional policies. It is not simply a series of articles reviewing the variance in spending between and across districts in regions, states, or provinces or the formulas used to calculate those variances. Instead it s-t-r-e-t-c-h-e-s us as finance professionals. It calls on "voices from the field"— voices speaking not just about domestic tranquillity but about global educational environments. It focuses on technology, uses technology, and looks at the power of technology to improve education from within—and without—its boundaries.

In this volume, you will hear voices raised in song and laments of woe. The strength of this yearbook is its daring to go beyond where we have ventured before to include topics and people who work in education both inside and outside our universities. You will hear about wonderful new ways of training administrators and the institutional frustrations that accompany "innovation." You will hear voices of praise for the exciting possibilities technology has brought into our lives as researchers and policymakers and hear also the clamoring sounds made in resistance to change by students and colleagues alike. You will hear also duplicity at times—duplicity from state-level policymakers who want educational systems to embrace technology but don't themselves understand the personnel, training, or capital burdens such technological foci place on schools, their managers, or communities. You will hear the sounds of war—and of peace—from individuals who at great personal sacrifice to their own safety believe the future lies with children and that technology can bring them back from the devastational edge of man-made conflict. This is "storytelling" in the finest tradition—a tradition of rich history in the sociological field, now coming into its own in education finance. These are stories that will enrich us and provide fodder for our intellectual and professional growth.

In this volume, you will hear from individuals who are school finance professors, career diplomats, policy analysts, researchers, and administrator and teacher educators. But they are *not* primarily "techies"—that is critical for understanding what you will read. These are folks just like you who want the best for their students, colleagues, profession, and progeny. They are users of technology. They see its fiscal as well as professional and sociological implications manifest in everything they do each and every day. They are reporting on where we are and where we need to go to improve training and education within and across our borders and convince policymakers and legislators what types and quantities of resources we need to accomplish that vision.

I prefaced this book by saying it was a unique volume. This is, indeed, a "New Age" in the annals of education finance. I hope this is the dawning of a renaissance in which such voices will continue to be raised to enrich our profession and bring new energy and synergy to what we do and reverberate with how we plan to plan for the generations that come after us.

About the Contributors

James F. Angevine is a Research Specialized Consultant with the Education Committee of the Pennsylvania House of Representatives. Prior to joining the Education Committee, he served in public education for more than 21 years. He has been a school administrator for more than 16 of those years, with specializations in planning, finance, and technology. He is currently responsible for analyzing educational finance and technology policy issues; maintaining a K-12 district database; and working on other education issues, such as school vouchers and charter schools. He holds both the master's of science in education and the master's in education in school administration.

Hank Bromley is Assistant Professor in Sociology of Education and Associate Director of the Center for Educational Resources and Technologies at the State University of New York at Buffalo (UB). He received his PhD in educational policy studies from the University of Wisconsin—Madison. His interests center in the areas of education and social change, the politics of technology, and feminist theory. He has recently published in *Educational Theory* and in *Teaching Education* (with Alison Carr). His book *Education/Technology/Power: Educational Computing as a Social Practice* (co-edited with Michael W. Apple) is currently in press at SUNY Press.

Faith E. Crampton is Professional Associate in the Economics and Education Finance Program of the Research Division of the National Education Association, Washington, D.C. She holds a PhD in Educational Policy and Leadership with an emphasis in Public Finance from the Ohio State University. Prior to her position with NEA, she was Program Principal in Education Finance at the National Conference of State Legislatures where she was primary author of *Principles of a Sound State School Finance System*. She has also served on the Graduate Faculties of the Schools of Education at the University of Oregon and the University of Rochester (New York). Her research has focused on state funding of education to achieve greater equity and efficiency. She has published widely in journals such as the *Journal of Education Finance*, the *Journal of School Business Management*, and the *Journal of the Council of Education Facilities Planners International*. She is Past President of the Fiscal Issues, Policy, and Education Finance Special Interest Group of the American Educational Research Association and has served on the Board of Directors of the American Education Finance Association.

Srebren Dizdar studied English and French literature at the University of Sarajevo (1974-1978, BA in 1978), with postgraduate studies in literature at University of Belgrade (1978-1979) and University of Leicester (Great Britain, 1979-1980). He received his MA in modern English and American literature from the University of Leicester, 1981, and his PhD from the University of Sarajevo, 1989. He was a Fulbright Scholar at the University of Southern California, 1987–1988. From 1982 until 1990, he served as journalist, international relations coordinator, Chief of Staff, and media officer for Radio and television Sarajevo, and from 1990–1992, he was Editor-in-Chief for a biweekly "Business Newspapers" in Sarajevo. He has served as Assistant Minister for Science in the Ministry of Education, Science, Culture, and Sports of Bosnia and Herzegovina (1993); as Permanent Secretary in the Ministry of Education, Science, Culture, and Sports (1993-1996); and, currently, as President of the UNESCO National Commission for Bosnia and Herzegovina. He serves as Assistant Professor at the University of Sarajevo Department of English (since 1994), where he teaches 18th- and 19th-century English novels and English romantic poetry; and in the Department of Comparative and World Literature (also since 1994), where he teaches the 20th-century novel. He writes articles on English, American, and African literature, as well as education, history, politics, and media, and translates poetry and fiction from English into Bosnian.

Karen Fullerton, MEd, is a doctoral candidate in instructional design and technology and a research associate with the GINIE and PENDOR projects at the University of Pittsburgh School of Education. Her research interests include television effects and computer interface design. She has taught technology and communication courses at the University of Pittsburgh, Duquesne University, and the Pennsylvania State University (Beaver).

Marianne Handler is Chair, Technology in Education Program, and Associate Professor in the Technology of Education Program, National College of Education, National-Louis University. She has been working with students, teachers, and technology since the mid-1980s. Her research interests center in the areas of technology integration in the curriculum, including supports needed by K-12 practitioners and reforms for providing technology experiences for students in pre-service education. She has been using the ISTE Foundation Standards for Technology as a model for designing technology experiences in teacher education courses. She is coauthor of *The Data Collector* (Intellimation), a Hypercard-based tool for analyzing qualitative data. She is coauthoring (with Ann Dana) the second edition of *Hypermedia as a Student Tool: A Guide for Teachers* (Libraries Unlimited/Teacher Ideas Press).

Stephen L. Jacobson is Associate Professor in the Graduate School of Education at the State University of New York at Buffalo (UB). He received his PhD from Cornell University, and his dissertation was the recipient of the American Education Finance Association Jean Flanigan Award for Outstanding Research in the Field of Educational Finance. From 1992 to 1997, he was Coordinator of UB's Educational Administration program and is currently Director of the Center for Continuing Professional Education. He also helped to design and instruct the Leadership Initiative for Tomorrow's Schools (LIFTS), a new approach to the preparation of school leaders. His research

interests include the reform of educational administration preparation and practice, teacher compensation, and absence and the social
organization of the school workplace. In addition to numerous journal
articles, his most recent books include *School Administration: Persistent
Dilemmas in Preparation and Practice* (Praeger, 1996) and *Transforming
Schools and Schools of Education: A New Vision for Preparing Educators*
(Corwin, 1998). In 1994, he was awarded the University Council for
Educational Administration Jack Culbertson Award for outstanding
contributions to the field of educational administration by a junior
faculty member.

Barbara Y. LaCost, Associate Professor in the Department of Educational Administration at the University of Nebraska—Lincoln, has
past experience as a public school teacher and coordinator in special
education and federally supported programs. Her research interests
include equity analyses of state funding plans, implementation of
state fiscal policies, and issues related to allocation of resources at the
school site. She has published in the *Economics of Education Review,
Journal of Education Finance, Educational Considerations,* and the *Journal
of School Leadership.* She was lead author of the resource allocation
chapter in the National Policy Board for Education Administration's
Principals for Changing Schools. Currently a member of the AEFA Board
of Directors, she also serves as Vice-Chair of the School Finance Research Committee for the Association of School Business Officials and
as Chair of the Fiscal Issues, Policy, and Education Finance SIG of the
American Education Research Association. In the past year, she has
been instrumental in the development and delivery of courses offered
to master's students through the use of LotusNotes Groupware.

Maureen W. McClure is Associate Professor at the University of
Pittsburgh and Director of the Global Information Networks in Education (GINIE) Project. GINIE provides information services to those
working in education for humanitarian and technical assistance and
economic partnerships in nations in crisis and rapid transition to
democracies and market economies.

F. Howard Nelson is Senior Associate Director of the Department of
Research of the American Federation of Teachers (AFT) and has been

with the AFT since 1986. He received the master's in economics and PhD in educational policy studies from the University of Wisconsin—Madison. Prior to joining the department, he taught school finance in the Department of Educational Policy Studies at the University of Illinois at Chicago. He has published many articles in journals such as *American Education Research Journal,* the *Journal of Law and Education, School Business Affairs, Education and Evaluation of Policy Analysis, National Tax Journal,* the *Journal of Education Finance,* and others. In addition to producing AFT's annual 50-state salary survey, his recent work includes an analysis of the private management of public schools in Baltimore and a report on the enrollment and costs of the Cleveland voucher program. Recent reports are posted at http://www.aft.org/research.

Alan T. Seagren has served as public school teacher and administrator, college faculty, Director of Summer Session, Associate Vice Chancellor for Academic Affairs at the University of Nebraska—Lincoln, and Vice President for Administration for the University of Nebraska system. In 1992, he returned to full-time teaching as Professor of Educational Administration, specializing in higher education. He created the Center for the Study of Higher and Postsecondary Education, for which he serves as Director. His early research focused on staff development training using videotape and observational systems for feedback. Recently, his research has focused on departmental leadership, and he has coauthored three books related to the position of chairman. Internationally, he has been a visiting scholar in Australia, Japan, China, and Scandinavia. He has held elected office in four national organizations and is a member of the board of directors for several community agencies and church units. He has been a staff member for the Institute for Leadership Development conducted by the National Community College Chair Academy and most recently has provided the leadership for delivery of a distributed doctoral program in educational leadership using LotusNotes Groupware with students in several states in the United States, Guam, Canada, and Australia.

Sheldon L. Stick is Professor in the Department of Educational Administration at the University of Nebraska—Lincoln and works primarily in the Center for the Study of Higher and Postsecondary Education. He has conducted on-line graduate-level computer courses for the past 3 years using LotusNotes Groupware. He coordinates 7 to 10 sections of the Foundations of Modern Education Program for the Teachers College each semester and participates in the team teaching of an experimental section of the program. He has been an elementary teacher, basketball coach, high school debate coach, and military officer and served on the faculty of three land grant postsecondary institutions and held adjunct professorships at two other institutions. He has been coordinator of a clinical service program, department chair, and campuswide multidisciplinary training and service program director. He holds a Certificate of Clinical Competence in Speech-Language Pathology and worked as a private practitioner and consultant for 17 years.

Emily Vargas-Baron is Deputy Assistant Administrator of the U.S. Agency for International Development (USAID) and Director of the Center for Human Capacity Development, responsible for the agency's global programs in education and training. She received her PhD in anthropology from Stanford University and was an Associate of the Stanford International Development Education Center. For 4 years, she worked in UNESCO as an educational planner in Europe, Latin America, the Caribbean, and the Middle East. Subsequently, she served for 6 years with the Ford Foundation as the Education Advisor for the Andean Region of Latin America. She conducted research projects in Colombia, including the first national-level study of nonformal education. On returning to the United States, she taught at the University of Texas. She founded and directed the Center for Development, Education and Nutrition (CEDEN), located in Austin, which has five service centers and many program replication sites in Texas and other states. Concurrently, she was a consultant for USAID, nongovernmental organizations, state and federal agencies, regional organizations, and national governments. She is the author of many books, articles, and research reports on educational policy planning, nonformal education, early childhood development, higher education, and computer-based educational technologies.

Cecilia Wandiga is a Policy Analyst for the city of Pittsburgh's Planning Department. She performs research and analysis of a variety of subjects ranging from demographic changes to employment strategies for specific populations. She received the MS in public management at Carnegie Mellon University. Some of her accomplishments include mention in the 1997 *Who's Who in the East*, presenting at the 1995 Midwest and Northeastern Regional Meeting of the Comparative and International Education Society, and designing a database currently being used by the Commission for Workforce Excellence to match employment opportunities with available training programs as well as determine training needs for Allegheny County. She has lived in Connecticut, Colorado, Puerto Rico, and Kenya and hopes to live to see the worldwide eradication of poverty.

Michael L. Waugh is Associate Professor of Curriculum and Instruction at the University of Illinois at Champaign—Urbana. He has been involved with electronic networking since 1986. His primary research interest is in trying to establish the impact and long-term value of instructional interactions on electronic networks. A particular area of interest is examining how computer networking can be integrated into teacher education programs to facilitate apprenticeship experiences between students, teachers, student teachers, professors, private sector businesspersons, and other individuals participating in the educational process. He has most recently been exploring models of fostering teaching teleapprenticeships among undergraduate and graduate students with practicing K-12 teachers and students to enhance university teacher preparation.

Kathleen C. Westbrook is Science and Technology Associate and Master Trainer, Division of Educational Programs/Educational Technology, Argonne National Laboratory—East, Argonne, IL, and co-owner of the DeerStar Group, Ltd., a technology research, development, and training corporation. She conducts training for the Chicagoland Educational Networking Consortium, local K-12 school districts, and Illinois service centers in the Chicagoland area. She is former Assistant Professor of Educational Leadership and Policy Studies, Loyola

University of Chicago and Portland State University, Portland, OR. She serves as Adjunct Professor of Education for National College of Education/National-Lewis University and Aurora University, Aurora, IL. She was awarded the PhD in educational administration with emphasis in financing educational technology and facilities from the University of Illinois at Champaign—Urbana. Her current interests center on electronic transactions and taxation of electronic commerce as a funding stream for schools, use of Groupware as a collaborative sharing tool for educational planning, and interactive CD-ROMs for human resource development service delivery.

Policy Changes Facing the Global Village

ONE

The GINIE Project
RESPONSIBLE NEIGHBORS,
MANY GENERATIONS

MAUREEN W. McCLURE

Networks to Reduce Expensive Isolation

Many school districts are isolated from each other and from their communities. Teachers often know little about what their neighbors in the community and in the profession are doing, sometimes even when those neighbors are in the same building. Such isolation can be expensive when it leads to unnecessary duplication and waste. The Internet offers new opportunities for those in education to learn from their professional and community counterparts, whether they are 12 or 12,000 miles away. Networks for shared learning about education are increasingly important because many education communities in the world are struggling with similar problems and can benefit from each other's experience. Education community networks are not only a effective way to form human capital but also both a growing social responsibility and a generative source for the economy. Civil democracies, market economies, and complex technologies all require a

3

well-informed public to make individual decisions with critical consequences for many others, including future generations. As gaps between rich and poor continue to widen, low-cost access to high-quality, timely educational information, expertise, and telecommunications technologies will play an increasingly central role in education for humanitarian aid and economic development in the United States and internationally. The Global Information Networks in Education (GINIE) project is one response to these concerns. It is a long-term educational commitment to the protection of children's futures in communities under chronic economic stress.

The GINIE project is headquartered in the Institute for International Studies in Education (IISE), School of Education, University of Pittsburgh. GINIE is sponsored by the Human Capacity Development Center at the U.S. Agency for International Development (USAID) in partnerships both with U.S. and international educational organizations. Its primary partnership is with an interagency consultation on education for humanitarian assistance sponsored by the United Nations Educational Scientific and Cultural Organization (UNESCO) International Bureau of Education (IBE), the International Council of Voluntary Agencies (ICVA), the United Nations High Commission for Refugees (UNHCR), the United Nations Children's Fund (UNICEF), and the United Nations Department of Humanitarian Affairs (UNDHA). The consultation's office is housed in the Education for Humanitarian Assistance Unit of the IBE in Geneva and is headed by Gonzalo Retamal, currently on a long-term mission in Iraq.

A Strategy of Responsible Neighbors

GINIE is a rapid response strategy for education in nations in crisis and rapid transition to democracies or market economies. Its mission is to provide rapid access to information and expertise related to humanitarian aid, technical assistance, and economic partnership for those working in education in these nations. It helps link people concerned about education in these countries with each other and with counterparts internationally. It accomplishes these goals through (a) an Internet repository of existing information and expertise that can be shared both on- and off-line; (b) complex, multilevel commu-

nications networks (phone, fax, e-mail, lists, on-line conferences); and
(c) programs of research, technical assistance, and training. GINIE's
intent is to (a) promote the formation of human capital through
international networks for professionals and communities by lowering information and transactions costs for educational improvement
and (b) demonstrate to the general public the ongoing contributions
of education to humanitarian aid, technical assistance, and economic
development.

GINIE's long-term interest is educational improvement that leads
to community and regional development, not only internationally but
domestically as well. It considers a community or regional education
sector as a whole, along with its related social service and commercial
links, as an *education economy*. An education economy grows out of a
response to both public and private needs for learning and innovation
across generations. The GINIE strategy does not seek to coordinate
these efforts through a centralized bureaucracy; rather it provides
opportunities for professionals to leverage scarce resources more
efficiently. This approach focuses on helping institutions and professionals lower their decision costs by providing access to information
and expertise that can help them "map" their position relative to
others under constantly changing conditions. For example, if potential donors and investors can "see" what others are doing, they may
be able to make better decisions about leveraging their own scarce
resources.

The GINIE strategy treats education as a "supramarket" activity. It
asserts that the relationships across children, parents, and teachers are
too complex to be captured by a metaphor of commodity markets of
supply and demand, driven by private property rights. It argues that
the consequences of today's educational relationships for current and
future generations cannot be adequately captured by contemporary
measures of return to personal or national income. Education is
deeply linked to more than income measures. It is necessary for the
legitimacy of states and economies. Democracies and market economies assume personal political and economic responsibilities for
sustainability because they are necessary for participation in self-
governance and markets. The public duty to protect children's futures
cannot be relinquished either in the voting booth or in the marketplace.

The consequences of this generation's investments in education, its finance, and its technology for the world's children will be experienced for many generations. GINIE challenges education professionals throughout the world to begin thinking about their responsibilities for children of the next millennium. Most education reform efforts in the United States and internationally focus on changing systems, not on sustaining civil and creative relationships over time. Reformers need to ask how education professionals, both in the United States and internationally, intend to contribute to the well-being of the world's current and future children. GINIE's longer-term relationship approach extends reform thinking to include a "dialogue of duty" for the next 50 generations.

Global Shared Resourcefulness

The GINIE strategy intends to avoid relationships of donor-recipient dependence by encouraging a longer-term "neighborly" approach that emphasizes cooperative self-interest across education professional communities both internationally and in the United States. It uses Inter/Intra/Extranet networks to help educational communities stay in close contact and keep each other informed as they respond to nations in crisis and rapid transition. The establishment of longer-term GINIE networks of communication for "shared resourcefulness" also helps contribute to the prevention of future crises through shared lessons learned.

GINIE's structure rests on a repository and on complex sets of networks that focus on (a) narrow professional themes such as land mine awareness education or improving educational quality at the classroom level under conditions of chronic economic stress and (b) country casebooks that share and analyze education efforts. GINIE's meta-network helps create a "one-stop shopping" site for those interested in education in nations in crisis and rapid transition. For example, communities in crisis may need rapid access to information and expertise about planning and financing in emergency conditions. Later, they may need access to materials and expertise about planning and financing recovery, reconstruction, and economic renewal. Teachers may also need rapid access to materials and expertise about methods such as active learning that can help traumatized children regain their resiliency. Later, teachers may need access to information and expertise about helping children and communities overcome the lingering effects of trauma.

Education Economies and Rapid Growth of Professional Capital

The core of GINIE's strategy of shared resourcefulness is a belief that the academic colleague metaphor for professional development and the responsible neighbor metaphor for community development can work together to make good economic and political sense in communities in crisis and chronic economic stress. Professional experience is encouraged to be widely disseminated at relatively low costs. By lowering information and transactions costs through shared expertise, professional capital, like knowledge capital in the academy, can form quickly, thus creating and renewing an information infrastructure for the profession. Educational improvement can be enhanced more rapidly if professionals offer to share pro bono some of their work and experiences rather than remain isolated.

The neighbor metaphor is also helpful for shared development across education communities. Sometimes neighbors need a hand up, sometimes they need advice, and sometimes they have something to share or to trade. Neighbors are involved for the longer term. Only by working together can they preserve and enhance their property, keep their communities safe from bullies, and create a responsible generational legacy. Good neighbors share tools and experience, giving something with the expectation that something as good or better may come back to them someday. GINIE focuses on the intersections of professional and school communities in the hopes of creating more efficient communications for neighbors so they are less isolated and more innovative.

The Children's Generals

Strategy is too complex to be modeled. It plays out in stories. Here is one GINIE story; there are others. Apollo-Ridge School District, a rust belt community nestled along the bucolic Kiskiminetas River in rural western Pennsylvania, struggled with lost local industry, the costs of environmental renewal, and an expensive and aging telecommunications infrastructure. The resulting economic problems are not unique to western Pennsylvania, as parents in many parts of the world worry that their children may have to leave home to find work.

With more wit than money, the community of Apollo generated innovative ways of leveraging scarce community and school resources to help children learn about development. Bill Kerr, the school district superintendent, directed many local development activities. He asked students to serve on the Apollo Area Economic Development Committee. Middle school teachers designed state-of-the-art, active learning, interdisciplinary curricula based on Apollo's revitalization project surrounding the town's old canal system. The tiny Apollo Trust Company, under the leadership of Ray Muth, became the first bank in the United States to act as a low-cost Internet provider for the community, creating free access to children and residents who kept a checking account with the bank. This small community bank also won the international top corporate banking Web site award in 1996.

Shortly after the Dayton Accords, Emily Vargas-Baron at USAID and Rob Fuderich, the senior education officer for UNICEF/Bosnia and Herzegovina (BiH), were concerned that "surviving the peace would be as difficult as surviving the war." UNICEF sponsored a trip to the United States for the author of those words, Srebren Dizdar, then Permanent Secretary for the federation Ministry of Education, Culture, and Sports in BiH. The visit to Apollo was organized by the GINIE project at the University of Pittsburgh. Dizdar had worked tirelessly under extraordinary conditions to keep children and teachers safe and connected with each other in "war schools." A network of education professionals, the "children's generals," spent the war years, mostly without pay, risking their lives daily to preserve education continuity for the next generation. These real-life heroes were remarkably successful.

Dizdar met with Apollo's community leaders and schoolchildren. He told them that Bosnians were grateful for U.S. sacrifices to end the war and ensure a just peace. Now that the war had ended, Bosnians were concerned about what would happen to them "after the CNN camera lights go out." He said the country needed to rebuild quickly so that U.S. troops could return home to their own families.

Back to the Future

Dizdar said Bosnians were headed back to the future. They faced three problems, all at the same time. First, the economy was operating at 7% of its prewar capacity. It would take a long time to fully recover from the war. He said teachers joked that they had been on a "4-year sabbatical from the world" during the war and needed to catch up on developments in education that they had missed. He said he hoped that GINIE would use telecommunications technology to help Bosnians create domestic and international networks of educational colleagues and neighbors who would be interested in sharing information, expertise, and dialogue about new developments in education.

He said the international and Bosnian communities had few resources to help local communities make the transition from the war and reconstruction to decentralization and economic trade. He hoped that education exchanges could help ease these painful transitions. He invited both Apollo community leaders and the Secretary of Education for Pennsylvania, Eugene Hickok, to think not only about educational aid and assistance but also about initiating trade delegations. He suggested that not only would a system of teacher and student and policy counterpart exchanges be good for children and professionals, but credible education networks would also be good for business as well.

The second problem Bosnians faced was that the Dayton Accords mandated that the Bosnian educational system be devolved to the cantonal (county) level. The Bosnian educational system was highly rated before the war, but it was a very centralized national system in the former Yugoslavia. This rapid devolution created serious planning problems, as there was very little resident expertise in education administration or finance. New cantonal ministries were being formed, and there were highly experienced professionals in education, but few with the skills and experience needed to plan for decentralization. He said he came to Apollo to learn how communities in the United States worked closely with their schools in a decentralized tax collection system. The United States has one of the most decentralized educational systems in the world, so U.S. professionals had a lot of experience with helping the public understand how local schools contributed to local communities. Dizdar hoped that a small network of education administration and finance professionals might be willing to share some time from their busy schedules to act as mentors in exchanges with Bosnian counterparts.

A third problem was that the introduction of new telecommunications technology created education opportunities that were not available before the war. Bosnians wanted to take advantage of these new technologies to "leapfrog" into an educational system that prepared children to work in a global economy and to build a sustainable domestic democracy. Dizdar hoped that children in the United States and in BiH would use the Internet to "grow up together" in this new technological environment, teaching each other now, trading with each other in the future.

He said Bosnians needed and wanted partnerships, not handouts. They needed longer-term, ongoing counterpart networks to exchange educational information and expertise. He said that Bosnian, Croatian, and Serbian teachers, students, and townspeople, like their counterparts in the United States, were trying hard to protect their children's futures by providing them with the best education possible. Like parents in Apollo, they too needed to develop their communities so that their children would not have to leave home to find work.

Need for Longer-Term Commitments From
Regional Institutions

During the war, school buildings were devastated, and many teachers fled the country or were killed. Groups of committed teachers and parents founded war schools and created safe havens for children to learn. Many teachers took great risks in their personal lives for 4 years to ensure that children did not lose a single year of schooling. In Mostar, for example, where there was heavy fighting, one elementary school was shelled more than 50 times, sometimes for sport. Defiant teachers helped children travel to and from school throughout the war and not miss a single year of schooling. Djulsa Bajramovic, the dean of the Pedagogical Academy in Mostar, lost her home, her husband, her only son, and later her dog. She stayed on to provide leadership for the war schools. Their record? Amazingly, not a single casualty for students traveling to and from school.

Strong networks of teachers were formed to share the materials they created and the lessons they learned. Clever teachers continually invented clever new ways to help students learn under extreme conditions. These creative networks held the system together during the war and became the core for the reconstruction of the educational system aftr it. Teachers during and after requested training in active learning methods from UNICEF so they could help children too traumatized by the war to be able to return effectively to more traditional pedagogies. Seth Spaulding at the University of Pittsburgh responded to these requests by establishing a field office in BiH headed by Lynn Cohen. Later, a team of University of Pittsburgh faculty members traveled to BiH to provide training workshops throughout the winter after the Dayton Accords, often without heat, electricity, or water. Instead of investing their time and efforts in earning consulting fees in safe, sunny climates, these professors skidded across icy roads with land mines close by.

Dizdar told the people of Apollo that a long-term commitment to mutual responsibility and respect was necessary to rebuild the country. He said that long-term, stable democratic relationships and market economies rested on the development of mutually respectful exchanges. He said that teachers and students need a deep continuity of support in their lives. Most project-oriented assistance could not provide it. School communities and professionals in other countries could. He asked if he could return with other Bosnian educators to visit Apollo and other school districts to learn more about public education and community partnerships in the United States.

In October 1996, the University of Pittsburgh professor Noreen Garman, codirector of the Institute for International Studies in Education and a seasoned traveler across icy Bosnian back roads, headed a technical assistance project with the Bosnian Ministries of Education for participatory planning for the renewal of teacher education programs in BiH, sponsored by the World Bank and the U.S. Information Agency (USIA). GINIE was an implementing partner. Now the president of the UNESCO National Commission for Bosnia and Herzegovina and a University of Sarajevo professor, Srebren Dizdar led a team of 14 Bosnian federation Muslim and Croat educators to Pittsburgh and Washington, D.C. The team consisted of many of the war schools' children's generals. For almost all of them, it was their first trip out of BiH since the war. Many of these cosmopolitan European professionals (one had been a diplomat before the war) had been unable to leave their communities because they were under siege for 4 years. During the war, all of them spent long periods of personal terror without the comforts of electricity, heat, water, and sometimes food.

Apollo was their first experience with intact schools similar to their own before the war. The contrast between the devastation in their own communities and the scrubbed schools of Apollo was very difficult to observe, even for real-life heroes. Mira Melo walked into the school library and said softly, fighting back a tear, "Ours used to look just like this." Apollo teachers, students, and community members rose to the occasion, showering the team with warmth, sensitivity, and compassion. Soon, frivolity emerged over a pickup game of basketball in the gym.

At the closing ceremonies for their visit, the Bosnian team, seated on stage, asked if anyone in the auditorium had served or had family members who were serving or who had served with the U.S. military in BiH. A few people in the audience stood. The Bosnian team rose to their feet and applauded them, no longer fighting back the tears in their eyes.

Humanitarian Aid, Technical Assistance, and Economic Partnership

The GINIE project encourages the development of networks of educational counterparts to exchange information and expertise about issues they have in common. It uses technology as a base for linking education professionals and school communities through humanitarian aid, technical assistance, and economic partnership. The strategy asserts that (a) education professional and community relationships, where possible, should be long term, civil, and creative; (b) educational relationships should emphasize shared responsibility, self-help, and development for each side; and (c) aid, assistance, and partnership relationships should be considered together as an ongoing national and subnational regional effort for cooperation across the education sector.

U.S. education efforts for aid, assistance, and partnership activities internationally are often initiated and performed by different well-intentioned people and agencies, nonprofit organizations, schools, colleges, and universities, often with little coherence or continuity. On one hand, the U.S. economy can no longer absorb the waste created by these isolated institutional efforts for aid, assistance, and partnership. On the other hand, a centralized coordinating bureaucracy would only exacerbate the inefficiency created by overfragmentation. The GINIE strategy of shared resourcefulness at the community level suggests that flexible electronic networks of rapid access to information, materials, and expertise, structured with regional education economy hubs centered in research universities may offer one way of lowering the costs of institutional planning, implementation, and evaluation.

Sic Transit Marshall

The GINIE approach to development complements more traditional approaches or "big aid." The post-World War II Marshall Plan approach infuses large cash flows into a postcrisis economy to help it build momentum. Unfortunately, many of today's crisis regions have economies that are too weak to respond adequately to a Marshall Plan approach of an intense quick boost to the economy through reconstruction as public works projects. Most of these regions faced conditions of chronic-to-severe economic stress before the crisis, making it very difficult for postcrisis reconstruction efforts to build and sustain political and economic momentum. In the past, the Marshall Plan helped restore legitimate economies that helped support legitimate states. All too often today, crisis creates illegitimate economies that gather momentum during the war and persist into a fragile peace.

These illegitimate economies profit from war and have little to gain from a stable state, a legitimate economy, and peace. Today's at-risk and crisis regions often have poor tax collection systems and weak militaries that can neither contain nor control these "gang" economies. Gang economies are sometimes so privatized that they protect themselves through the use of paramilitary or mercenary troops, many of them with equipment and training superior to those of the local police or even the nation. Even more troubling, the new crisis economies are not found only in developing countries. Pockets of gang economies now erode the quality of education and threaten children's futures in communities in developed countries as well.

This wrenching disjuncture between a weak legitimate public state and a powerful, privatized illegitimate economy requires new responses more effective than the post-World War II Marshall Plan alone. How can today's forces for legitimate governments and markets overcome or subvert the momentum of illegitimate gang economies? What credibility will education have to a child with no access to a legitimate economy? In the United States and internationally, much of the education reform movement is focused on improving educational efficiency. Increasingly, in regions in crisis, rapid transition, and chronic economic stress, however, educational reform is a powerful and painful struggle for political and economic legitimacy.

GINIE' s presence offers a kinder, gentler approach to complement Marshall Plan efforts through providing and sustaining a low-profile, long-term international presence across transitional polities and economies.

*Bosnian Professionals Have Much to
Teach U.S. Counterparts*

One of the great lessons Bosnian teachers on all sides of the war have taught the rest of the world is that basic education is more than teaching literacy in school buildings. It is more than a traditional civil service job with a small but steady paycheck. It is more than a new building or a winning athletic team. It is a fundamental moral commitment to the protection of children's futures. This protection lies not in buildings or government mandates but in the civility and resourcefulness of people who chose not to allow their ethnic identities to overwhelm their professional ethics.

Much of the educational reform rhetoric in the United States focuses on the financing of systems. The debate is important, but it too often diverts attention from the core of teaching: the ongoing civil and creative relationships that teachers, students, and community members make with each other. All else is a prop, useful but less essential. Bosnian teachers, faced with the loss of school buildings, taught in basements. Faced with the loss of electricity, water, and heat, they taught with flashlights, creating books with special paper that could be used in dim light. Teachers in Gorazde, faced with violent sieges aimed to disrupt schools and civil life, gathered children in the streets after a shelling to sing songs and play music. The real heroics of Bosnian teachers remind the rest of the world of what it too often forgets, that children are precious gifts that civil societies must protect.

These moral commitments to "zones of peace" are public and cannot be privatized. Democracies and free markets rest on the assumption that people are sufficiently well educated to make informed decisions at the polls and in the markets. A nation's continuing well-being rests on the civility and creativity of its citizens. Civility is not genetic. Each generation must learn it anew. It must be modeled by both parents and teachers. Education reform is as much about relationships of dignity, respect, and innovation as it is about systems of bricks, budgets, and buses.

Implications for U.S. Education Reform

The current rush to privatization may unleash consequences that will be regretted. In a world where birth control privatizes the child-bearing decision, the rationale for investment in children's education can turn ugly. Parents want what is best for their children, but what about the children of others? The most rational decision for parents is to maximize the welfare of their own children. In a world of scarce resources and competition, that welfare can be maximized in two ways. The first is by enhancing their own children's access to educational resources. The second is by minimizing the access of competitors.

Why should parents invest in educational opportunities for their own children's future competitors? Liberals traditionally counter that children in a school district or a state "belong" to it, and so the wealth of the district or state as a whole should drive education funding. Conservatives traditionally imply that those who chose to have children should bear the primary responsibility for their education, so parents need to have as much control over education subsidies as possible to make the best choices.

Highly decentralized financial systems funded primarily through residential property taxes tie children's educational opportunities to the wealth of their parents and not to the wealth of the nation as a whole. Is there a tacit abdication of personal responsibility for a public duty to future generations of children embedded in current reform arguments? Is there a growing popular cultural acceptance of the notion that the privatization of childbearing means that children can be treated as durable goods? Is the direction of policy that one may choose either to buy a refrigerator or to have a child? The purchase and maintenance of a refrigerator is not subject to government subsidy, so why should those who choose not to have children be taxed, since those who choose not to buy refrigerators are not taxed? This argument is echoed quietly in many suburban communities: "Why should we pay for those who choose to have children they cannot afford to educate?"

The GINIE strategy resists the commodity argument, claiming that children cannot be equated with consumer goods. Children are supramarket. One has relationships with children, not with refrigerators. The moral duties of one generation to the next cannot be traded away. Each generation of children must learn how to successfully inherit or create and maintain both civil democracies and responsible market economies. Refrigerators do not vote.

The GINIE strategy argues that learning is primarily a public event because its consequences are relational and thus not easily privatized through property rights. Children, not refrigerators, grow up to be drug dealers. Children, not refrigerators, renew their economies. Refrigerators do not construct or maintain legitimate governments, outwit criminal gangs, or preserve the ecology.

In Sum

The GINIE project is a bold new educational development strategy that connects educational counterparts internationally through professional and community networks. GINIE is concerned about sustaining high-quality teaching internationally, especially in nations in crisis and rapid transition to democracies and market economies. The project promotes the idea that educational systems rest on networks of responsible cross-generational relationships of teaching and learning. It asserts that education is more than a commodity to be traded in the marketplace—it is a public duty to distant generations. The GINIE strategy poses the following tenets:

1. Education is a moral commitment to the protection of children's futures.
2. The education community (teachers, parents, students, and others) need to publicly demonstrate their contributions to public civility and economic innovation.
3. Education professionals bear a special responsibility to share generously information and expertise with their educational neighbors.

Resources

Those who are interested in sharing their resourcefulness with others are invited to contact the GINIE project through the Internet at www.pitt.edu/~ginie, by e-mail at ginie+@pitt.edu, by phone at (412) 624-1775, by fax at (412) 624-2609, and by snail mail at GINIE c/o IISE, 5KO1 Forbes Quad, University of Pittsburgh, PA 15260, USA.

TWO

State Financing of International Higher Education Partnerships for Trade and Development

EMILY VARGAS-BARON

Increasingly, state policymakers are becoming aware that for their states to compete successfully in the international marketplace of goods, technologies, knowledge, and ideas the funding of international higher education linkages and networks is an essential part of their states' investment strategies. Because the future of many U.S. communities is being shaped by the quality, appropriateness, and timeliness of local responses to global market forces, states are seeking to expand and improve their education systems; to develop market-driven workforce training programs; and to forge international alliances both for trade and for helping to attain sustainable social, environmental, and economic development in far-flung nations.

AUTHOR'S NOTE: This chapter does not necessarily reflect policies of the United States Agency for International Development.

International higher education partnerships include both "linkages" and "networks" that are formed to conduct common strategies to attain shared objectives. Linkages include exchange agreements between one or more U.S. institutions of higher education and one or more such institutions in foreign countries. They are often formed in collaboration with other public agencies and private sector organizations. Networks include collaborations between existing networks of higher education associations in the United States and in other nations or regions. Both of these major forms of international higher education partnerships have yielded positive results in increasing international trade, creating markets, and attaining sustainable development goals.

The Changing International Arena

As a nation, we are moving from a one-way, largely federally financed system of development assistance to two-way systems of development cooperation—of development partnerships. During the past 30 years, the vast preponderance of federal support for international development has been devoted to providing a variety of one-way aid activities, including technical assistance, training, and financing or credit programs. Little if anything was expected in return, and host nations rarely structured their programs for mutuality. In addition, earlier development assistance efforts were composed largely of federal initiatives, and relatively little state or private funding was sought or expected.

Nations in Latin America, Asia, the Middle East, and now in Sub-Saharan Africa are requesting—and demanding—to play a greater role in donor relations and coordination. They are seeking to develop active reciprocal exchange relationships with industrialized nations. Today, we are witnessing the first stage of truly global strategies for development cooperation that are built on reciprocal exchange relationships. As a nation, it is evident we are moving away from a unitary system of bilateral assistance toward shared systems of collaboration and cooperation for building development partnerships. The various creative roles of state governments and private sector institutions in international development and trade are becoming increasingly apparent.

Higher education institutions, including 2- and 4-year colleges, public and private universities, and research institutes, represent a rich resource of international programs, faculty, students, and alumni who can help to forge new international partnerships. States are becoming aware of key capacities found in institutions of higher education. State legislators and administrators are including those institutions in statewide strategies for international trade and development. Such state-level strategies usually encompass the following elements:

- Market analyses regarding industrial capacity and plans
- Resource analyses that include the higher education sector
- Plans to improve state visibility through proactive marketing of state resources
- Financial incentives and tools for promoting exports and trade
- Decentralized coordination of higher education and public and private sector organizations

Complementing state-level investments in international development through sponsoring the international programs of higher education institutions are the growing investments by city and county governments, professional associations, chambers of commerce, corporations, businesses, financial houses, and foundations. New systems of funding consortia are arising, often with federal funding playing an initial catalytic role through helping institutions to establish and nurture valuable partnerships and through providing limited seed money that is complemented by state and other sources. Businesses, corporations, and financial houses can also encourage state legislatures and agencies to invest in international higher education partnerships that they are convinced will promote trade, development, research, and training opportunities.

In the future, a significant proportion of the foreign policy of the United States will be based on, or influenced by, the complex web of international collaborations that institutions and associations of higher education are able to build. We in the United States have observed repeatedly that one-way "charity" can lead to the rejection of the giver. Solid two-way ties of mutual interest and common positive experiences tend to bind the members of reciprocal exchange relationships, however. Higher education institutions in many nations seek to create partnerships with U.S. colleges and universities. In addition, many other industrialized nations are sponsoring such linkages as proactive ways to build their international marketing and trade. Several multilateral organizations are assisting industrialized nations to forge these new alliances. It is of strategic importance for U.S. institutions of higher education to collaborate with such linkages and to forge new ones rapidly that are focused on major development and trade issues. Major global issues include food security, basic education and workforce skills training, reduction of fertility rates, disease control, environmental protection, democratic governance, crisis prevention, emerging markets, economic growth, and international trade. Institutions of higher education in the United States and abroad must collaborate on joint action research and development programs. Special attention should be given to overcoming major development problems, which often are serious barriers to expanding markets and international trade.

Persons from developing nations who have been trained in the United States as a part of development assistance activities or through scholarship and exchange programs represent an important reservoir of talent and contacts for creating partnerships with the higher education institutions they attended. Many of these U.S.-trained specialists are now the leaders of their nations, directing major public and private institutions. Often, they are eager to build long-term exchange relationships with their former alma mater and other universities and colleges in the United States

Selected Case Studies

Higher education institutions and associations increasingly are playing a crucial role in forging international partnerships for trade and development. Various striking examples of international higher education partnerships with state support or assistance now exist. Three higher education institutions and one association program have been selected to illustrate the main points of this chapter. The Arizona case study illustrates the utility of both municipal and state involvement. The Florida example represents a comprehensive state planning effort that includes higher education partnerships. The Washington state program demonstrates how "globalizing" university programs positively impacts economic development and international trade in a state.

Frequently, these international partnerships include one or more U.S. institutions of higher education; federal support from the U.S. Agency for International Development (USAID) or another federal agency, such as the U.S. Department of Agriculture (USDA) or the U.S. Department of Commerce; one or more state agencies; a municipal government; one or more private businesses, corporations, or associations; and one or more foundations or professional associations. When well structured, these partnerships generate considerable local enthusiasm and many activities that yield results far exceeding those of the original program design. At the state level, initiatives have included the institutional reform and development of higher education institutions; the establishment of distance learning/virtual university programs; the creation of innovative workforce training programs; the provision of investments in economic, infrastructural, and institutional development; the expansion of tourism; the creation of regional development initiatives; and even the signing of trade agreements between states and developing nations.

For a variety of reasons, some states are beginning to forge what amount to virtual foreign policies for marketing and trade in the global arena. Increasingly, these state-level initiatives should be linked with federal-level policy planning for development cooperation, global marketing, trade development, and workforce training for employment, both at home and abroad.

Arizona:
Maricopa Community College District (MCCD)

The Maricopa Community College District (MCCD), a large community college system serving 18 communities and more than 177,000 students in Arizona, has boldly created multifaceted alliances called "Partners in Development" with Arizona state and municipal agencies, corporations, and hotel chains. With partial funding support from USAID through the Association Liaison Office for University Cooperation in Development (ALO), MCCD has established thriving partnerships with a variety of institutions in Mexico and China. As a land-locked state, Arizona is making a significant effort to enter the Mexican and Pacific Rim markets, to maximize use of scarce water resources through promoting innovative irrigation programs and dry farming, and to foster active tourism.

MCCD developed a linkage with the Universidad Autonoma de Baja California Sur (UABCS) through the ALO and Mexico's National Association of Universities and Higher Education Institutions (ANUIES). They established a "long-term collaborative relationship in human resources training, international economic development and research between the two institutions, with their respective private sectors and states, to further economic development" (Landrum, 1997, p. 2).

The two focus areas selected initially were (a) the tourism and services industry, and (b) water resources management. On the U.S. side, the actors include the MCCD, State of Arizona Small Business Development Center Network, Greater Phoenix Economic Council, City of Scottsdale Economic Development Department, Maricopa Skill Center, Hyatt Hotels, Paradise Valley Community College, Scottsdale Community College, and Arizona University. The Mexican side included UABCS; State Secretary for Enterprise Development; State Economic Development and Promotion Agency, State Office of Tourism; the Los Cabos and the La Paz tourism industries, including many hotels and services; the National Water Commission; and municipal and state water commission representatives.

After an initial identification of problems of mutual concern, Arizona representatives visited Baja California and a return visit was made to Arizona. These visits included intensive discussions; site visits; and concrete planning activities including surveys, alliance formation, and assessments of training needs on both sides. In many instances, these visits constituted the first time that the governmental and private sector representatives met and considered common needs and in turn entered into discussions with education and training institutions responsible for technical training in Baja California. The results to date include the following:

- The first education, industry, and government partnerships in Baja California have been created, with UABCS joining the State Council of Tourism Development and the National Water Commission.
- UABCS has been invited to form a small business development and deregulation entity.
- Training programs have been planned in both states on low water use landscaping and water/wastewater treatment as well as hospitality programs, on how to do business in both countries, and on how to promote trade between the states.
- Future plans have been developed for linking industries, education, and government in the two states.
- Plans are proceeding for a feasibility study on water resource issues and needed educational support, a survey of employment and training needs in La Paz, further work on Los Cabos labor needs, and further exchange visits.
- Secondary schools in La Paz have agreed to add a course in tourism and water culture. (Landrum, 1997)

With regard to exchanges between the cities of Phoenix and Chengdu, China, MCCD has helped to design and conduct specialized training courses for 18 Chinese delegations averaging 12 persons each with the active participation of more than 118 Phoenix local businesses and government professionals in the training courses. Topics have covered many areas of federal, state, and city planning and services, as well as selected aspects of business development and trade (Landrum, 1995). In turn, Phoenix businesses have benefited from Chinese visits, learning about investment opportunities in China. Lacking a business office in China, the state of Arizona looks to the city of Phoenix and MCCD to help develop business and trade relationships in China, thereby supplementing state resources for international economic development.

Florida:
"Enterprise Florida" and the University of Florida

In 1992, the state of Florida moved aggressively to expand its economy, increase employment opportunities and training, and improve its status in international trade and marketing through the creation of "Enterprise Florida." This coalition includes the Florida

University System, led by the University of Florida with the active participation of the state's leading community colleges; corporations and businesses; federal laboratories; and state and municipal governments and their agencies.

Enterprise Florida created the Council of Business Leaders and the Technology Innovation Corporation. The University of Florida Office of International Studies and Programs (OISP) collaborates with Enterprise Florida and provides technical services to Florida's businesses and industries. Parallel to Enterprise Florida is the Florida Linkage Institute Program of the Florida International Affairs Council, whose avowed purpose is to create linkage programs between universities and community colleges with international business partners, the state of Florida and higher education institutions abroad.

Results have included the following:

- Creation of new companies and employment opportunities
- Increased state funding for international programs linked to trade, marketing, business development, and development
- A more international outlook and preparation on the part of the student bodies of Florida universities and community colleges and an increase in departmental and professorial capacity to deal with challenges of the global marketplace
- More than 11 linkages established with partial support from the state of Florida under support from the Florida Linkage Institute Program
- A network established by the Florida Linkage Institute Program and the ALO of Latin American universities with Florida universities mutually dedicated to improving the administration and management of higher education institutions in Latin America and Florida
- Businesses and corporations assisting institutions of higher education to gain increased funding from the state legislature because they now value the technical support they receive from higher education institutions

Washington:
Washington State University (WSU)

Washington State University (WSU) boasts one of the most highly developed systems of international higher education linkages in the United States. Guided by the vision of creating a "Global Land Grant

University," WSU has forged 12 regional nodes with more than 52 university linkages in more than 48 nations. This striking dedication to international development and trade reflects not only the service philosophy of this major state university but also Washington state's role as "the sixth largest exporting state in the U.S. after California, Texas, New York, Michigan and Illinois" (Murphy, 1996, p. 5). International trade now provides employment for one fifth of Washington state's workforce, which represents "nearly 600,000 jobs" (Murphy, 1996, p. 5). Washington state's leading trading partners include Pacific Rim nations, Canada, and the European Union (EU); WSU also maintains linkages with all other world regions.

Recognizing the importance of higher education partnerships for promoting trade and development, Washington state leaders decided to provide funding to WSU for its regional education, research, and economic development networks and the International Marketing Program for Agricultural Commodities and Trade (IMPACT). These programs were established to internationalize teaching, research, and public service, as well as to provide technical guidance and research support for agricultural associations, commissions, and private firms seeking to enter world markets.

WSU has embarked on an audacious program aimed at "internationalizing its teaching research and public service/outreach programs, forging strategic alliances with public institutions and the private sector throughout the world" (*The Image of the World*, 1995, p. 1). All aspects of the university are involved in this internationalization process, from administrators to faculty and alumni at home and abroad. In all partnerships, strategic alliances are sought with state agencies and the private sector. WSU has collaborated with the Asia/Pacific Economic Cooperation (APEC) and participates actively in several APEC working groups. It also has several grants and contracts with USAID and USDA, as well as with other federal agencies; nongovernmental, business, and community organizations; and regional agencies for Latin America, Africa, the Middle East, Europe, and Russia and the newly independent states.

The five main areas of collaboration present in most WSU partnerships are

- Policy and cooperative research
- Cooperative graduate education programs

- Student and professorial exchange
- Distance education, with Internet virtual university programs
- International alumni programs

WSU has contributed actively to policy dialogue with representatives of state agencies and leaders in other nations to promote marketing and trade. As a result of its presence in such dialogues and its technical resources on agricultural and marketing matters, WSU has also played key roles in assisting state agencies with the negotiation of important international trade agreements. The university has assisted private firms and state agencies to understand foreign marketplaces and to enter into new trade relationships.

James B. Henson, the director of WSU international programs, recently commented, "Because of our orientation and interaction with the private sector, support within the state has been good." For WSU, extensive private sector involvement in higher education partnerships has been an important key to securing strong state support for WSU's global approach. Henson stated,

> Our state legislators and state government officials recognize the importance of what we are doing and are supportive of our programs. The university routinely participates in joint state-private sector trade missions and international visits. As we have greater successes and have opportunities for educating our public officials, we believe that support will continue to grow. These various activities comprise what can be termed an educational program, providing opportunities for input by a whole range of stakeholders in both the public and private sectors. They enhance private sector performance in the short, medium and long-term, while building capacity in U.S. education to respond to future needs. (Personal communication, May 1997)

New Educational Technologies: Global Virtual Universities

The Internet is providing U.S. universities and colleges, including community colleges, with the unique opportunity to provide, at low cost, extensive higher education services of excellent quality with higher education institutions abroad.

Each of the programs outlined above conducts distance higher education activities and maintains Internet web sites. In some instances, the very topic of the exchanges is learning technologies that can be used to support trade and development programs, as in the cases of Florida and Washington state. WSU provides full international university programs through interactive Internet programs, thereby enabling students to change campuses and still complete a full course of studies. Currently, more than one million students are enrolled in virtual classrooms (Gubernik & Ebling, 1997). Because of the high demand for virtual university services on the part of developing nations, it is likely that in the near future many more higher education partnerships will be formed around distance learning programs that are linked with marketing, trade, and social development issues.

Technical Support:
The Association Liaison Office for
University Cooperation in Development (ALO)

To create cost-effective and efficient international higher education partnerships, institutions of higher education need more than funding support from their state governments and other public and private sector partners. They also require excellent technical guidance and contacts to establish fully successful international partnerships that can be sustained until they meet all of their key objectives. The Association Liaison Office for University Cooperation in Development (ALO) was established in 1992 to coordinate international activities of six major higher education associations and their memberships of more than 2,400 institutions with USAID. The ALO provides direct services aimed at assisting the U.S. higher education community to build international partnership programs. Since 1995, the ALO has conducted five Higher Education Roundtables on development challenges, with a special focus on establishing higher education partnerships (ALO, 1995, 1996a, 1996b, 1997a, 1997b).

In addition, the ALO has led the development of three international networks and, in turn, several higher education linkages. These include a network with selected U.S. universities and community colleges and the Mexican National Association of Universities and Institutions of Higher Education (ANUIES) in Mexico, a network with the Florida University System and several Latin American universities, and a network between universities of the U.S. Southeast and the Carpathian region. Based on proposals from the higher education community, more networks and linkages for economic and social development will be established in the future. The ALO seeks to work closely with state governments to fulfill its mandate to develop more thriving higher education linkages and networks. Because the ALO constitutes a valuable resource for state planners and policymakers, contact information is provided at the end of the chapter.

Lessons Learned

To ensure they become important players in global educational, economic, and social development, states should channel a growing portion of state agency budgets to supporting the establishment and maintenance of proactive, innovative international higher education linkages and networks for trade and development. Through policy dialogue, states that decide to create international partnerships should establish a collaborative state policy, strategy, and plan for identifying, supporting, constituting, and nurturing strong higher education linkages and networks focused clearly on establishing international trade and sustainable development activities. Although it is relatively easy to create higher education linkages from higher education networks, it is very difficult to create a higher education network from discrete linkages not forged through an existing network. States should consider helping to establish international higher education networks as well as linkages.

State-level and private sector support can be secured subsequent to initial funding from a federal source, but to ensure comprehensive programming and sustainability it is preferable for state funding to be provided from the outset. This will help also to encourage higher education collaborations to emphasize attaining the economic and social development results that are sought by the state. Once states witness positive yields from their investments in international higher education partnerships, they tend to expand their support for such

activities. By beginning a few successful higher education partnerships, state agencies will be able to demonstrate their utility and to
expand their number over time.

Some Steps for Developing Successful
International Higher Education Partnerships

Experience reveals that to develop successful international higher
education partnerships, the following steps should be pursued.

1. From the outset, representatives should participate in all planning, implementation, and evaluation processes from (a) all government levels (federal, state, and municipal), (b) private and service
industries, (c) labor unions and professional associations, and (d)
higher education and training institutions (community colleges, public and private universities, technical training schools, and secondary
schools). These institutions should maintain their independence, but
they should seek to build new, balanced, reciprocal relationships.

2. A shared vision and shared expectations and objectives must be
identified, reviewed, and reinforced frequently. Relationships of trust
must develop from this process or the partnerships will founder.

3. The benefits for all parties must be identified, reviewed, and
achieved to the extent possible. Both the U.S. and developing nation
institutions must achieve clear and tangible benefits in all phases of
partnership activities.

4. State funding, agency and legislative representation, and lines of
communication should be established from the outset, if possible. It
is advisable to establish a state council for international higher education partnerships for trade and development that has representatives
from all sectors listed above. The council should be empowered to
help create new partnerships, guide existing ones, and disseminate
information about its activities. In some instances, the council could
be given the responsibility for managing a partnership development
fund created by the legislature, municipal governments, corporations,
and other private sources. This fund would provide grants and contracts to proposed and existing partnerships. In other instances, specific state agencies or existing associations may be given the respon-

sibility of managing such a partnership development fund. The fund should include counterpart support from other governmental and private sources.

5. Face-to-face exchange visits and meetings must be carried out periodically in each locale, and they should always involve the active participation of state and local decision makers. It is not enough to exchange e-mail messages or to conduct teleconferences. Personal relationships lead to interpersonal commitments and the achievement of shared goals.

6. A concrete partnership development program must be established by all parties. At a minimum, it should include key strategic priorities in the following areas:

- Trade, marketing, and social development activities
- Education and training programs in the state and abroad
- Focused research, assessments, or studies on key aspects of the collaboration
- Course development on topics of mutual interest
- Provisions for shared internal and external evaluation and monitoring

7. The action plan for the partnership development program should have clear action steps, list all responsible parties, and provide deadlines for the completion of activities. At the same time, it is essential for the program to maintain flexibility, ensure the timely execution of activities, and retain the capacity to revise plans in light of emerging opportunities or problems.

8. Periodically, all partners must assess their own and others' participation and achievements. All partners must use these assessments to help them plan their future action steps. By sharing the roles of evaluation and accountability, the parties are more likely to search together to achieve efficiencies and effectiveness.

9. Additional partners may join existing partnerships to round out institutional capacities. For example, this type of flexible development cooperation may help to unite land grant colleges with community colleges and private universities for partnerships to meet resource needs in other nations and to spread learning opportunities across the faculty and students of several campuses. This complexity will require the fast-paced prioritization of activities and the development of

timely response systems. Virtual universities and Internet linkages will assist with the management of these mushrooming partnership systems. State governments that begin to harness these far-flung systems may well develop more competitive systems for increasing state trade and economic and social development.

What Are Some Benefits of State Higher Education Partnerships for Trade and Development?

Active engagement in international development through higher education linkages and networks helps to promote the "deparochialization" of state governments and enterprises. International partnerships help states to forge successful long-term international agreements for mutual development. The participation of higher education institutions assists all involved to promote intercultural understanding and to develop the knowledge and skills that humanize relationships for trade and development. State administrators, businesspeople, and professors who engage in developing higher education relationships usually report they feel more effective in conducting international relations. Higher education partnerships positively affect many faculty, administrators, and students, thereby helping them to become responsible and culturally competent actors in global trade and development.

Higher education partnerships will assist states to achieve increased economic growth and international trade and marketing. The social and economic development programs that result from higher education partnerships will assist less developed nations to prosper and to achieve the status of "emerging markets."

Conclusion

Fundamentally, as with traditional societies, reciprocal relations at the international level can form a solid base for promoting socially responsible and economically successful market exchange. Increasingly, institutions of higher education are becoming leaders in forging the groundwork for foreign policy, development cooperation, and international trade. Because of the globalization of the U.S. economy, it is influenced negatively by crises, violence, and economic disruption in developing nations. With more than 33 nations in crisis as of this writing, and a fourfold growth in expenditures for humanitarian and military assistance, the avoidance of national and regional warfare is essential. Through creating strong economic and social ties for mutual development by improving and expanding basic education and skills, many of the requisite conditions for long-term stability and sustainable development will be established. The higher education community is an essential "motor" for the creation of innovative partnerships for trade and development.

Increasingly, leaders in state governments are becoming aware that for their economies to progress, they must look to resources in the higher education sector to help them increase trade and compete successfully in the global marketplace. Wherever possible, it is advisable for states to promote strong partnership activities for trade and development. It is in the strategic global interest both of the United States as a nation and of the individual states to engage in cooperative activities with institutions of higher education devoted to promoting international trade and development.

References

Association Liaison Office for University Cooperation in Development. (1995, December). *The look of development cooperation ten years out: What new roles for the state, higher education, business and industry, and the community?* Policy Roundtable Series: Higher Education and Global Development, Association Liaison Office for University Cooperation in Development, USAID.

Association Liaison Office for University Cooperation in Development. (1996a, June). *The greying of development expertise: What's needed and how will the next generation get trained?* Policy Roundtable Series: Higher Education and Global Development, Association Liaison Office for University Cooperation in Development, USAID.

Association Liaison Office for University Cooperation in Development. (1996b, March). *Higher education, the corporate sector, states and communities: Forming new partnerships for economic development.* Policy Roundtable Series: Higher Education and Global Development, Association Liaison Office for University Cooperation in Development, USAID.

34 POLICY CHANGES FACING THE GLOBAL VILLAGE

Association Liaison Office for University Cooperation in Development. (1997a, April). *Increasing the relevance of higher education to development: What U.S. and Mexican public/private partnerships can do.* Policy Roundtable Series: Higher Education and Global Development, Association Liaison Office for University Cooperation in Development, USAID.

Association Liaison Office for University Cooperation in Development. (1997b, September). *Regional roundtable: USAID/Higher Education Partnership in Development.* USAID.

Gubernik, L., & Ebling, A. (1997, June). I got my degree through e-mail. *Forbes,* 84-92.

The image of the world. (1995, November). Washington State University, Pullman.

Landrum, B. (1995, June). "Chengdu—Phoenix International Economic Development Professional Training Program." Report prepared for the Maricopa Community College District.

Landrum, B. (1997, May). *Maricopa Community College District, Arizona and Universidad Autonoma de Baja California Sur, Mexico.* Report prepared for the Maricopa Community College District and the Association Liaison Office for University Cooperation in Development.

Murphy, L. (1996, July). *Washington state trade picture.* Washington Council on International Trade.

Resources

Publications

Building bridges between institutions of higher education in the Carpathian region of East Central Europe and the Southeastern United States. (1996, December). Report of Phase Two the Cooperative Pilot Network Activity, Associated Colleges of the South and the Association Liaison Office for University Cooperation in Development, USAID.

A changing university for a changing world: Michigan state's global future. (1995, February). A report by the International Review Committee, Michigan State University, East Lansing.

Henson, J. B., Noel, J. C., Gillard-Byers, T. E., & Ingle, M. D. (1991, May). *International universities: A preliminary summary of a national study* (Occasional Paper No. 7). International Programs, Washington State University, Pullman.

Lowe, E. (1996, March). *Higher education in the Americas: The role of universities in hemispheric development.* Office of International Studies and Programs, University of Florida, Gainesville.

Pallan Figueroa, C., Claffey, J. M., & Adelman, A. (Eds.). (1995). *The relevance of higher education to development.* Coleccion Biblioteca de la Educacion Superior, Asociacion Nacional de Universidades e Instituciones de Educacion Superior, Institute of International Education and Association Liaison Office for University Cooperation in Development, Mexico, D.F.

Contacts

For additional information on international higher education partnerships for trade and development contact:

Dr. Joan Claffey
Association Liaison Office for University Cooperation and Development (ALO)
One Dupont Circle, N.W.
Suite 700
Washington, DC 20036
Telephone: (202) 857-1827
Fax: (202) 296-5819

Internet: alo@aascu.nche.edu

The ALO was established by six leading higher education associations:

The American Council on Education (ACE)
The American Association of Community Colleges (AACC)
The American Association of State Colleges and Universities (AASCU)
The Association of American Universities (AAU)
The National Association of Independent Colleges and Universities (NAICU)
The National Association of State Universities and Land-Grant Colleges (NASULGC)

THREE

Education, Technology, and the Social/Economic Order

SREBREN DIZDAR

CECILIA WANDIGA

This chapter examines research that illustrates the link between education, technology, and economic growth. Discussion centers around
unanswerable questions and issues in the hopes of both promoting
debates and focusing future research not on solutions but on the
dilemmas and complexities arising with the dawn of a new age.

The Social/Economic Order of Things

If by some means we could create a tabula rasa for all countries
today, thereby making them all equal, what would enable some
countries to advance faster than others? Thus far, there have been four
main ways or theories as to how this would come about: developmen-
talism, evolutionism, economics, and education.

Developmentalism:	Evolutionarism:
A country's advancement is determined by its rate of natural progression through a predetermined set of stages.	A country's advancement is guided by the rate of proliferation of a trait(s) that increases adaptability to the existing environment.
E.g.: crawl-walk-run agrarian-industrial-post industrial	E.g.: the use of technology to modify one's surroundings seems to dominate in colder climates, where the consequences of not having such a trait are much more immediate
Economics:	Education:
A country advances according to the use/allocation of its resources. E.g.: the right combination of capital investment, natural resources, human capital, trade regulations, fiscal policy, etc.	A country's advancement is dependent on its ability to develop an educational system that generates and maximizes intergenerational wealth transfer. E.g.: the transfer of technical/production know-how, entrepreneurial skills, applied technological know-how, etc.

Figure 3.1. Theoretical Paradigms for Economic Growth

After World War II, it became necessary to rebuild the great powers of Europe and repair the damages done to Japan. This issue of how to best and most effectively promote development in a country created the field of development economics. Initially, developmental assistance came exclusively in the form of capital (either monetary or machinery) transfers. The main focus of relief efforts was to replace what had been lost during war. Around 1960, the European powers found themselves largely restored, albeit without the majority of their former colonies. Additionally, there were now two new forces in the arena of global supremacy: the United States and the Soviet Union. The birth of the United Nations had also imposed a new civility on the world order, and it was no longer proper etiquette to dominate the world through colonialism.

During the Cold War, the world once again divided into allies and foes, with allies being either prodemocracy or procommunism. Since the economically advanced countries were on a relatively similar footing in strength, they were left with no choice but to turn to the less developed countries to expand their "spheres of influence." The key to this expansion was hidden under the guise of humanitarian relief, and the field of economic development was popularized. Through the popularization of humanitarian relief, a "holy crusade" aimed at spreading the knowledge and resources from the developed countries to all other nations. Almost 40 years later the developmental holy war can also be deemed a failure.

With the exception of a few converts such as the TIGER[1] countries in Southeast Asia, the rest of the underdeveloped world remains largely or comparatively underdeveloped. Granted, there have been many improvements, most notably in areas such as health. Additionally, even underdeveloped countries have experienced growth, many times significant growth. For example, according to the 1995 World Bank World Development Report, Nepal experienced an average annual growth rate in gross domestic product (GDP) of 2% from 1970 to 1980 and 5% from 1980 to 1993; China experienced a growth of 5.5% from 1970 to 1980 and 9.6% from 1980 to 1993; yet they are classified as low-income economies, despite the fact that all the high-income economies, with the exception of Australia, the United Kingdom, and Switzerland, experienced a decline in average annual GDP growth over the same period. Unfortunately, this growth has failed to translate into the transition from poverty to wealth for the majority of underdeveloped countries.

Today's answers to the question of what makes some countries advance faster than others are (a) their ability to develop and adapt new technology (note that this was the explanation originally offered by evolutionarism), and (b) their ability to educate their workers on how to develop and adapt such technology (education theory revisited). Before accepting these theories at face value, we must examine whether there is any evidence to substantiate either of these approaches.

A Brief View of Current Research

The Effects of Education

Although there is a vast body of research on the issue of what makes an education system effective,[2] there is very little research that proves a direct link between improvement in education and economic growth, let alone development (see Smelser & Swedberg, 1994, pp. 581-199).

When examining any system, it is important to consider both the *intra-effects* and the *inter-effects* of its subcomponents. The intra-effects are any changes (or lack of change) to the overall system caused by various parts of a system. For example, any education system has the following components: finance/spending, classroom size, student-teacher ratio, subject matter, teaching methodology, number of classroom hours, administrative personnel, equipment and materials, and educational attainment. Consequently, any examination of what makes an education system effective thereby focuses on the intra-effects, for example, what effect does increased spending have on educational attainment as measured by SAT scores?

The Economist (1997) compared intra-effects in educational systems across countries. The research assumed that the economies were a result, at least in part, of the educational system—that is, causality was a foregone conclusion. This may or may not be the case. For example, U.S. per pupil spending is more than three times as much as that of Hungary, and yet Hungary outperformed the United States on both the math and science Third International Math and Science Study (TIMSS) tests (*The Economist*, 1997, p. 21). If one were to follow the logic of the underlying assumption, one would expect to find that Hungary's economy outperforms that of the United States because its education system is superior. The 1995 World Development Report, however, classified Hungary as a middle-income economy and the United States as a high-income economy. This is one indication that the strength of a country's education system may not have a direct bearing on its economic status. Such results are disturbing when one considers that current education reform rhetoric in most First World countries promulgates the assumption that a top ranking in educational performance tests will lead to a top ranking in economic performance tests. The reality is that a myriad of factors affect the status of an economy, and education is just one of those factors.

In light of this, we must then consider the inter-effects of education on the economy. The inter-effects are any changes the system or its subcomponents cause on other systems, for example, the effect of educational attainment on economic growth. Research conducted by Richard Rubinson and Irene Brown shows the following:

1. Increases in enrollment lead to increases in economic growth when the increases in enrollment are the result of a demand for new or improved skills but not when the changes in enrollment are the result of social measures aimed at reducing class biases or differences, for example, affirmative action policies.

2. The quality of schooling, when controlled for time-on-task (the amount of hours spent learning), does matter for economic growth but only to the extent that there is a direct link between the type of training and the skills needed in the economy.

3. The effects of education on economic growth are a function of the link between the level of schooling provided and the sectoral distribution of the economy. For example, if an economy is dominated by low-skilled, labor-intensive sectors, then schooling aimed at producing such skills will increase economic growth, whereas other types of schooling will not; in other words, the economy determines the type of schooling that should be provided.

4. Cross-national studies of education show that there is much more similarity among educational systems than one would expect given the corresponding sectoral variations in the economies studied; therefore, a plausible conclusion is that political processes dominate the effects of education on the economy in that they impede the necessary feedback, from the economy to educational systems, that would enable greater variation to occur. (Smelser & Swedberg, 1994, pp. 594-596)

Thus we can see that (a) education does affect economic growth, and (b) the important issue is not to find out which economies have the better schools but rather to determine under what conditions education does affect economic output. The question that remains is, If it is true for all cases that for economic growth to occur the economy must dictate the type of schooling provided, how can a country create the type of economy it desires and simultaneously train its labor force without sacrificing economic growth?

The Effects of Technology

Depending on how technology is defined, there is either a little or a lot of evidence to show that increased technology leads to increased economic growth. If one limits one's definition of technology to a particular machine or revolutionary production approach, technology's contribution to economic growth then becomes limited to the longevity or usefulness of that particular machine or production mechanism. For example, the spinning wheel, although largely instrumental in England's economic growth during the 14th to 16th centuries, is absolutely irrelevant to England's economic growth today. But if one defines technology as the "totality of methods rationally arrived at and having absolute efficiency (for a given stage of development) in every field of human activity" (Ellul, 1964, Note to the Reader), then the evidence can be traced back as far as the history of humankind. In the field of economics, Adam Smith is credited with first noticing that productivity and efficiency increased through the division and specialization of labor. In other words, if we input humans into an economic equation with the variable name "human capital," their ability to produce will depend on "the totality of methods rationally arrived at and having absolute efficiency (for a given stage of development)."

Robert Reich (1992) defines technology not as a particular piece of machinery or a revolutionary technique but rather as the ability to develop "the skills and insights necessary to *continue* to invent" (p. 152) in an efficient manner. Technology as thus defined affects the rate at which goods can be produced in a given economy, and this in turn affects the rate at which an economy grows (assuming, of course, that the goods being produced are in demand and can be traded). Ellul (1964) also wrote,

> When technical progress intervenes, it modifies not only the application of economic laws but also the essence of the laws. We may consider this in two ways. First, economic laws are not eternal like the laws received by Moses on Mt. Sinai. Our economic laws are valid only for a certain type or form of economy. When technical progress occurs, it is integrated into the economic system not as a foreign element but organically. Technical progress is a part of the essence of an economic system, not a mere accidental event. . . . When the facts change, the constants as well as the laws, are modified. . . . In fact, technique has modified the scale of human economy, and the laws that held for the average economic system at the beginning of the nineteenth century no longer apply in the new scale of the economy we know today. . . . Just as technique breaks down the barriers between economic sectors, so an economy based on technique tends to burst asunder the traditional sociological frameworks. (pp. 204-205)

In his (1995) book *The End of Work,* Jeremy Rifkin illustrates how in the 1920s technology had the following three effects: (a) a technological dilemma—rapid development of labor-saving techniques—very similar to that which we are facing today with the profusion of computerization; (b) increased productivity, which affected the economy as a double-edged sword—on the one hand increasing productivity and output while on the other sometimes reducing the number of jobs; and (c) mass production, which created a need for mass consumption—thereby reinforcing Ellul's (1964) statement, "Just as technique breaks down the barriers between economic sectors, so an economy based on technique tends to burst asunder the traditional sociological frameworks" (p. 205). The question then becomes, What effect is technology having today?

The Effects of Globalization

Most economists would agree that computer and telecommunications technology has pushed modern society to the brink of another industrial revolution. Although the exact nature of this revolution remains nebulous, or at least best defined as *globalization,* its effects are already being felt, and many more effects are anticipated. The effects already visible include the following:

- *Unprecedented speed in the area of communications.* The technological revolutions taking place with fiber optics, computers, and software now enable us to transmit data and graphic information faster than most of us can imagine. For example, Texas Instruments (TI) makes the following claim: "TI also is a leading supplier of asynchronous transfer mode (ATM) technology, a sophisticated switching technology that can route streams of multimedia traffic at mind-boggling speeds (it will be able to transmit a 3½-hour movie, such as *Gone With the Wind,* in.84 seconds)" (Texas Instruments, 1995).
- *A shift from the importance of ownership of natural resources to the importance of efficiently developing natural resources.* In *The Work of Nations,* Robert Reich (1992) points out, "The increasing dominance of Japanese automakers within the United States is due largely to the fact that they have been able to utilize American workers to make a higher-quality car, in less time than can American-owned automakers. Although the 1990 *Consumer Reports* ranks most Japanese cars higher in quality than American cars, it finds no difference

between the quality of Japanese cars produced in the United States and the quality of Japanese cars made in Japan" (p. 146).

• *Unprecedented speculation in world financial markets (largely enabled by high speed communications).* According to John C. Edmunds (1996), "Securitization—the issuance of high-quality bonds and stocks—has become the most powerful engine of wealth creation in today's world economy. Financial securities have grown to the point that they are now worth more than a year's worldwide output of goods and services, and soon they will be worth more than two years' output. Historically, manufacturing, exporting, and direct investment produced prosperity through income creation. Wealth was created when a portion of income was diverted from consumption to investment in buildings, machinery, and technological change. Societies accumulated wealth slowly over generations. Now many societies, and indeed the entire world, have learned how to create wealth directly. . . . The total dollar value of all investment-grade securities worldwide that could potentially be issued is upward of $150 trillion, roughly five times the value of annual world output. The value of all quoted securities was $35 trillion at the end of 1992. This value will approach $60 trillion by late 1996 and surpass $83 trillion by the year 2000. Therefore, only 40 per cent of these securities have been issued so far. More are being issued every business day, and their value on the whole is increasing" (pp. 118, 120).

• *The emergence of a new player in world power—financial rating institutions.* Previously, power lay in the hands of those with either economic or military might, but financial rating institutions are now beginning to use their ratings as a leveraging tool to dictate the actions of governments. It should be alarming to education finance professionals that the authority behind national fiscal policy making is being transferred from elected bodies politic to private for-profit investors. In February 1995, Moody's Investing Service threatened the Canadian government with a lowering of its credit rating unless it adopted "tougher" fiscal policies (Wiens, 1995, p. 17). During the U.S. budget impasse in January 1996, matters appear to have been more expediently resolved after Moody's "threatened to downgrade the rating of $387 billion in U.S. bonds . . . [because] the impasse on the budget and debt had 'significantly increased the risk of a default'" (*U.S. News & World Report*, 1996).

• *Rising unemployment levels worldwide.* Just as electricity increased productivity and efficiency in the U.S. automobile industry during the 1920s, the computer is increasing productivity and efficiency today. Human capital is being replaced with machinery, and unemployment levels are rising. According to Jeremy Rifkin (1995), "In the Organization for Economic Cooperation and Development (OECD) countries, 35 million people are currently unemployed and an additional 15 million 'have either given up looking for work or unwillingly accepted a part time job.' In Latin America, urban unemployment is more than 8 percent. India and Pakistan are experiencing unemployment of more than 15 percent. Only a few East Asian nations have unemployment rates of below 3 percent" (p. 198).

• *A need to create "new" or* emerging *markets for the consumption of surplus goods.* One way of increasing economic growth is to find ways to increase consumption, because it can guarantee growth in production; consumption can be stimulated on the domestic front or abroad. During the colonial era, the colonizing countries (e.g., England, Spain, Portugal, France, etc.) extracted raw materials from their colonies (e.g., India, Kenya, the Philippines, Argentina, etc.); processed these materials into goods such as clothing, tea, and chocolate; and sold them back to the colonies or within their domestic borders. Today, saturated "domestic" markets with rising unemployment levels create the need for rapid economic growth elsewhere, thus making underdeveloped nations a prime target for the consumption of industrialized goods. The late U.S. Secretary of Commerce Ronald H. Brown made the following statement at the opening session of Big Emerging Markets (BEMs)[3] Conference in Washington, D.C., on July 24, 1995:

> The BEMs were experiencing rapid economic growth, and exports have been a major source of new high-wage jobs in the United States. In fact, since 1987, exports have contributed over one third of the overall growth in GDP, rather impressive since exports are only 12 percent of the economy. Export-related jobs have increased at six times the rate of domestic employment growth. Given the Clinton Administration commitment to restoring opportunity to all Americans, job creation is our chief objective. And exports clearly are the key. While trade with our traditional partners in Europe and Japan was still crucial, these are mature economies, growing slowly. So, we looked to emerging markets and found vitality. We found that in the largest of the emerging markets not only had the decade just past brought tremendous growth, but the decades ahead offered even greater promise. (Big Emerging Markets (BEMs) Home Page, 1995)

• *A collapse of the social welfare engine and, in response, the rise of nongovernmental organizations (NGOs).*

The independent sector [NGOs] is playing an increasingly important social role in nations around the world. People are creating new institutions at both the local and national levels to provide for needs that are not being met by either the marketplace or public sector. . . . England's experience is closest to that of the United States: it has thousands of volunteer associations, . . . There are currently more than 350,000 voluntary organizations in the United Kingdom, with a total income in excess of £17 billion, or 4 percent of the gross national product. . . . In France, the third sector is just now beginning to emerge as a social force. In one recent year, more than 43,000 voluntary associations were created. Employment in the third sector has been growing lately, while jobs in the formal economy have been declining. The social economy now accounts for more than 6 percent of total employment in France, or as many jobs as are provided by the entire consumer-goods industry. (Rifkin, 1995, pp. 275-286)

Rifkin reports comparable figures for Germany, Italy, Japan, Eastern Europe, Central Europe and the former Soviet Union, Pakistan, Sri Lanka, South Korea, Bangladesh, Nepal, Indonesia, Korea, Peru, Brazil, Dominican Republic, Colombia, Mexico, Burkina Faso, Kenya, and Zaire.

• *Bi-modal economies.* The shrinking of the middle class in many countries across the globe is stirring fears of a return to feudalistic days during which there were only two classes—the filthy rich and the dirt poor. Robin Broad and John Cavanagh have observed,

> Thirty years ago, the income of the richest fifth of the world's population combined was 30 times greater than that of the poorest fifth. Today, the income gap is more than 60 times greater. Over this period the income of the richest 20 per cent grew from 70 to 85 per cent of the total world income, while the global share of the poorest 20 per cent fell from 2.3 to 1.4 per cent. The number of billionaires grew dramatically over the past seven years, coinciding with the spread of free-market policies around the world. Between 1987 and 1994, the number more than doubled from 145 to 358. According to our calculations, those 358 billionaires are collectively worth some $762 billion, which is about the combined income of the world's poorest 2.5 billion people. At the bottom, 2.5 billion people—approximately 45 per cent of the world's population—eke out an existence using just under 4 per cent of the world's GNP. At the top, 358 individuals own the same per cent. (pp. 26-27)

• *Centralized versus decentralized government control.* In the face of the current social and economic transformations caused by globalized systems, governments are struggling to find the most effective way of maintaining control and allocating resources. In the United States, emphasis is being placed on centralization or nationalization. According to Lester Thurow (1992),

> With the collapse of much of its banking sector, by early 1991, the American government had been forced to take over two hundred billion dollars in private assets and was expected to end up owning three hundred billion dollars in private assets before the hemorrhaging stopped. A government corporation, the Resolution Trust Corporation, has become by far the largest owner of property in America. (p. 16)

Centralization is also happening in the field of education with the promulgation of national education standards led by the newly created National Center on Education and the Economy. Paradoxically, Europe is simultaneously attempting to centralize regional efforts through the creation of the European Union (EU) while decentralizing utility companies and government services

(England); educational systems (France); and, in the case of Eastern Europe, the entire economic structure.

• *Globalized education reform.* Over the past 10 years, governments have begun paying close attention to cross-national educational attainment ratings. Once results are released, virtually every industrialized country undertakes some level of educational reform aimed at maintaining its status or improving the next set of scores. The U.S., French, German, British, Hungarian, Japanese, and Swedish governments, to name a few, have all put pressures on their schools to improve teaching methods, lengthen hours, stress math and sciences, or implement any one or more of a myriad of policies. Underlying this reform effort are two beliefs: (a) "education is the key to getting rich—for countries as well as for individuals," and (b) as governments make "desperate attempts to rein in public spending . . . there is no prospect that governments will chuck money at schools without checking to see whether standards are improving" (*The Economist*, 1997, pp. 21-22).

From Research to Reality: Developing an Ideological Framework

We previously pointed out the difficulties involved with analyzing components of this highly complex system. We would, however, be remiss if we failed to illustrate the issues discussed in their proper contexts. The following three questions attempt to illustrate the shortcomings of current conventional education finance wisdom in order to redirect research and debate toward a more comprehensive approach. Given that we are currently in an age of transition, it is helpful to acknowledge the possible effects change may have on societies. Figure 3.2 below is a gross generalization of these effects.

The figure points out that as the rate of change accelerates, flexible societies tend to encourage reform measures as a form of adaptation, whereas rigid societies tend to encourage preservation of existing measures (or "tradition"). Also illustrated is the fact that flexible societies tend to prosper in stable environments (i.e., stable political, economic, and social structures), whereas rigid societies are frequently a reaction to crisis environments. Thus, in examining education finance policies and measures adopted in any given country, region, and so on, care must be taken to situate the given entity in the appropriate environmental context. Without this understanding, it becomes very easy to slip into a chastising mode (i.e., District X would

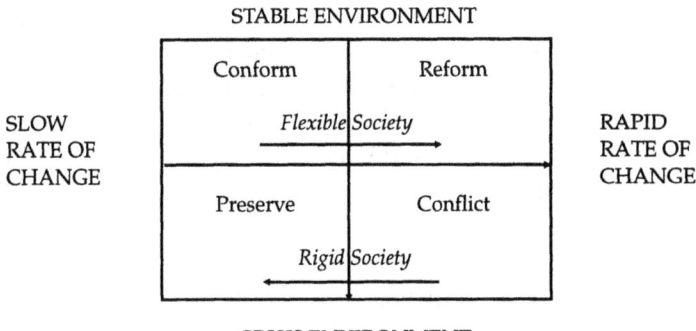

Figure 3.2. Anticipated Effects of Environment on Societal Change

achieve greater achievement if only it would implement policies such as those used by District Y) while forgetting that the behavior of the district in question is most likely the result of its particular environment and not just the district's actions or lack of actions. The purpose of this chapter is only to spark debate that reminds of us of our highly complex environment. We will, therefore, leave the details and examples of this model for future discussion and proceed with the examination of other key issues.

Who Pays? Who Controls? Who Benefits? Who Rates?

An overarching concern raised in complex transitional environments, particularly in light of the rise of nonprofit educational organizations and the reliance on securing debt as a means of accumulating wealth, is, where does the center of educational finance control or decision making lie? The social contract theory (in all of its varying forms) developed by Jean Jacques Rousseau, John Locke, and Thomas Hobbes maintains that individuals make a conscious and deliberate decision to surrender their individual freedom to the laws that govern a collective whole in the hopes of obtaining a more orderly and profitable (in the philosophical sense) existence. For example, even though we all could find plenty of alternative uses for our money, the majority of us pay taxes to a collective fund based on the belief that there are a number of necessary social goods (e.g., safety, education, roads) that make our existence more palatable but that we cannot afford to pay for as individuals. To better understand the complex influences on education finance policy in the United States, it will be important to examine post-Cold War influences on education finance in other parts of the world.

Education Finance as Social Control

Until now, the responsibility for administering these collective funds has been placed in the hands of governmental entities that, elected or not, have largely been able to appropriate their use in a relatively independent fashion as a mechanism for social control. As a result, the basis of governmental power and influence has largely rested on the fact that funding streams can make entities behave according to a prescribed modus operandi. For example, when the Pennsylvania Department of Education decided to establish a computerized link between all of its 401 school districts, a number of rural districts resisted the attempt, claiming that the network was "an artifact of the devil sent to corrupt the purity of their existence." After various attempts, the department finally succeeded in getting these districts to use the network by placing all budgetary information exclusively in electronic format. The ability of social control mechanisms to "fog the minds of men" is exercised by governments all over the globe. Education finance professionals cannot afford to lose sight of the importance of funding streams as a social control mechanism.

In European traditions, the state intervenes on behalf of an imaginary "public good" (the commonwealth). Depending on the level of democratic achievement, these funds are often used for a "good cause" (e.g., socialized health care and welfare benefits) or as a "bribe" against various competing interests (e.g., funds are now being used to buy a social peace in the EU and to direct changes in the former Soviet bloc countries). This intervention holds true in both Eastern and Western Europe.

In Western Europe, after the tragic experiment of fascism and Nazism, there were few attempts by educators to address issues of collective responsibility and funding. The emphasis shifted instead toward individualism. This emphasis on the alienation of individuals, so popular in post-World War II European existentialism and later in the U.S. "beat," "hippie," and "yuppie" cultures, masked critical issues of collective responsibility and public duty. Collective action was often limited to the labor and civil rights movements. This shift to individualism created a merit metaphor that was used to help shape education finance policy as a social control mechanism. For example, a post-World War II system of intermediary schools was introduced in Britain for the most talented individuals from working-class backgrounds. Traditionally, higher education had been reserved for the elite: Lawyers reproduced lawyers, doctors reproduced doctors, and so on. With the reform, a new kind of educated class from a working-class background infused fresh blood into the traditional professions and, eventually, strengthened them.

In the United States, the GI Bill was also used as a social control mechanism to help open up economic opportunities traditionally reserved for elites through limited access to higher education. The civil rights movement spawned a rising class of African Americans who made their way through the education system and became established as an example of "equal" opportunities within the white Anglo-Saxon Protestant (WASP) society.

In contradistinction to the Western European and U.S. obsession with individualism, Eastern European culture emphasized the collective. Education was proclaimed as the all-important tool or vehicle for conquering nature and an unjust past. Education, previously reserved for family elites, became an exclusive privilege granted party loyalists who were recruited from peasant and industrial worker backgrounds. Education was thus used as a way of strengthening party dominance over historical elites. The children of the former educated elites from bourgeois backgrounds were widely discouraged from studying in the country, so many fled abroad. The remaining found their way through the system with great difficulty. Many who dared to protest lost their good jobs and were assigned to dirty manual tasks (the current Czech president Havel suffered from this discrimination).

By the late 1960s, these tensions had somewhat eased. More universities opened with the sole aim of preventing further unemployment increases in Eastern Europe. At the same time, these countries borrowed heavily from the West to "catch up" with the developed world.

The result of living on debt was the collapse of industrial giants such as East Germany, Poland, and Czechoslovakia in the late 1980s when no more foreign loans were available to boost the rundown economy. The Soviet empire fell in large part because of its economic inefficiency. It also caused myriad social upheavals (first in Poland, then in Romania and Czechoslovakia and Hungary, and currently in Bulgaria and Albania). The education finance systems in each country underwent traumatic changes. There was little money to support growing numbers of students, the quality of teaching declined dramatically, and even if students graduated from a university there were few jobs available—not only prestigious jobs but any jobs. In the Ukraine, links between education and access to the legitimate economy were so badly damaged that many students refused to pursue higher education.

In some countries (Hungary, the Czech Republic, and to some degree Bulgaria), foreign interests, either governmental or private (such as the financier George Soros), wanted to see major changes in education finance policy. The communist system proved to be inflexible. It had always been traditional and centralist and was unable to reform itself. No longer were there state funds to provide necessary supports. The academic communities were ready neither to propose major changes to survive nor to redirect a course into the 21st century. Critical complicating matters included uneven international funding streams. Albania is a singular case in point. Albania almost received an influx of needed financial aid early in the 1990s to reconstruct its education system. Unfortunately, the Balkan war shifted the world political interest to Yugoslavia and, consequently, Bosnia and Herzegovina (BiH). Now, with Bosnia somewhat pacified, Albania has finally attracted attention, but only as a result of the latest unrest. Bloodshed and human suffering play a major role in determining international education finance policy.

Albania and BiH are small. Russia itself is a source for future unrest and conflict, and post-Deng's China may face the same difficulties. These situations are vital to U.S. interests, but they are not media-visible to its citizens. One education finance colleague reports the suppressed news of starvation among the elderly in Russia. Starvation is tough competition for taxes to pay for schooling.

Education in Eastern Europe collapsed overnight when there were no stable revenue sources. When the very foundations of revenues became too weak to support even the most basic needs, let alone services, people suddenly realized education was not the tool of progress. It does not provide for a good, or even decent, living any more. Many young people have lost faith in education as a fair opportunity to enter the legitimate economy. Democracy and market economies have brought hope to many; now it is up to education professional colleagues around the world to help each other learn how to live successfully within them.

Education Finance in a Retreating State

Today, as national fiscal policy-making authority shifts from political bodies to financial institutions, ratings, and stockholder whims and fancies, one cannot help but wonder, in light of the historical context provided, about the possible social consequences of this new style of decision making. Weak governments make weak regulations. Decentralization has been an efficient means for dividing governments and weakening their fiscal regulatory capacities. It seems as if we will now be forced to make hard-and-fast decisions on the old economic debate over what is best: government control or unfettered market action. Although the best alternative is probably a compromise between the two positions, the question still remains as to what the delineation of such a compromise would look like.

In other words, should resources be allocated according to "ability to pay" or "ability to benefit"? Or, if both ability to pay and ability to benefit are factored, what should be the proportion between the two? For example, should education be paid for through tax revenues (based on societal benefit) or corporate revenues (based on economic growth)? Or, if both are used, what portion of education should be paid for through taxes and what portion of education should be paid for through corporate revenues? In light of this, who determines curricula, those providing the funding or those performing the rating?

Furthermore, when government funding streams (or any other revenue source) collapse, what becomes of the systems they were supporting? Should laissez-faire policies be adopted in the hopes that these systems will rebuild themselves, or should some form of strategized intervention take place? Last, will private funding streams allow or enable the same degree of social equalization attempted by government funding, or will they only serve to augment the inequalities that already exist?

Education for High Technology Skills for All:
Our Salvation?

As economies become increasingly interlinked and countries struggle to maintain their competitive advantage under the pressures of globalization, heavy emphasis is placed on the development of high-tech skills among the general population. Left out of the equation are those such as artists, philosophers, social theorists, and believers in a broad educational base as essential for civility and innovation. Also overlooked is the fact that education may not be the only contributor to economic growth. In other words, it is not very difficult to conceive of situations in which substantial amounts of economic activity are being generated by individuals with low education levels, for example, professional athletes, celebrities, or criminals.

In light of this dilemma, schools are faced with determining how best to train students for tomorrow's society. It is our belief that the price of democracy is civility, and that civility means a well-rounded and broad-based educational background to manage modern complexity. Instead of learning subject matters as discrete entities, students should learn the interdisciplinary nature of the material as it is used by professionals. For example, to design a good building, an architect must draw upon the following fields: art (for the aesthetics of the building), structural design, physics (to determine the appropriate materials for the climate, etc.), math (for the calculations), sociology (for an understanding of how physical structures affect human behavior), history (to determine what cultural aspects the building will represent), geography (for terrain issues), horticulture (for landscape design), and business management (to oversee the project, control costs, etc.). Architecture is not the only profession requiring a multidisciplinary background. As a matter of fact, one could argue that all professions could benefit from the knowledge base of other subject matters. Unfortunately, just as obligatory simplification of pedagogy obfuscates the complexities of professional judgment, interdisciplinary links tend to get overlooked in the classroom environment. Education is more than the acquisition of skills for a "good job." It is an apprenticeship to civilization.

Our recommendation for schools is that classes should be structured in such a manner as to expose students to interdisciplinary linkages. This approach is not new and is used in many classrooms around the globe. The modern educational metaphor of machine efficiency as a standard of perfection creates "human computers," individuals capable of performing complex quantitative analyses but void of the ability to engage in moral/philosophical debates. We would like to encourage schools not to abandon the interdisciplinary approach that confronts both science and justice in the rapidly changing and globally interconnected society that is now emerging. The biggest source of educational intergenerational wealth transfer will not stem from a limited skills base (i.e., high tech) but from the social capital of civility and innovation. This requires a more thorough and complete understanding of each individual's and country's contribution to and place in the global infrastructure. The most daunting obstacle confronting schools and governments alike is finding a way to ensure that every individual receives an appropriate apprenticeship to civilization. Although we cannot control for individual aptitudes and ambitions, we must begin to find ways to provide everyone with a minimum set of social and economic skills before the increasing divide between rich and poor becomes so ineradicable that we give up on our long-standing hopes of poverty amelioration.

Globalization Versus the Decentralizing State:
Increasingly Fragile Boundaries

Another quandary emerging as communication speeds shorten the distances between countries and regions is how or where to allocate human capital. Truly liberalized trade calls for a free flow of all commodities, including humans. We have already seen how monetary resources can now be readily transferred across boundaries and the resulting effects of this in processes such as securitization. So too has the mobility of human capital affected education finance. On observing such a phenomenon, one would be inclined to think that traditional nation-state boundaries no longer matter. Nonetheless, they happen to remain firmly in place. Even among regional trading blocks such as the EU, there is still a strong resistance to uniform standards of language, currency, labor laws, and so on. Although countries seem quite willing to benefit from the positive terms of trade offered by globalized resources, their willingness stops when these resources begin to affect anything considered inherently nationally characteristic such as language, culture, and the like. Furthermore, there is an even stronger backlash when flows of human capital exceed some unidentified tolerable level. French hostility toward West African immigration has led to the warehousing of these immigrants in communes with conditions deplorable enough to attract the attention of the international news media. In Germany, hostility to immigrant groups such as the Turks has escalated to violent levels echoing less enlightened periods of the country's history.

Even a country built by immigrants such as the United States is adopting a tougher stance on its immigration policy. This occurs regardless of the skill levels of the groups that are immigrating and regardless of the economic benefit received from the activities of such groups. In the Spring 1996 issue of *Foreign Policy*, John Stremlau provides a case study of Bangalore, India, that highlights the debate behind this stance. Bangalore is rapidly becoming a software production center, rivaling and, according to the Capability Maturity Model developed by the Software Engineering Institute at Carnegie Mellon University, even surpassing most Western software production centers. Stremlau notes,

Citibank, American Express, General Electric, IBM, Reebok, Texas Instruments, Hewlett-Packard, AT&T's computer unit—NCR—and Compaq Computer are but a few of the many companies that depend on computer software developed and tailored to their needs in Bangalore and other Indian cities. . . . During the 1980s, Bangalore's main software-industry exports were not products, but people: high-skill, low-wage software engineers and programmers who took jobs in the United States, a process Indians call "body shopping." It is often said that India boasts the second-largest scientific community in the English speaking world—more than 3 million scientists and technicians. There are approximately 140,000 Indian scientists of all disciplines currently working abroad. Each year, the number of computer-science graduates from Indian universities averages around 15,000, and many go to the United States.

The hiring of foreign programmers by U.S. companies recently became a contentious topic in Congress. In 1995, legislation[4] was proposed that would significantly limit the number of American visas issued to foreign workers. . . . The export of semiskilled American jobs is not new. But thus far, American blue-collar workers have borne the brunt of these job losses. Workers in Bangalore, however, are not doing the routine assembly of circuit boards, as in Malaysia, nor are they doing only the mind-numbing work of data processing and minor software repair. Increasingly, Bangalore's engineers are designing the software programs that tell computers what to do. And American white-collar workers with annual salaries of $40,000 or $50,000 suddenly find themselves competing with equally well-trained and experienced computer programmers in Bangalore who often make less than $10,000 per year. (1996, pp. 152-168).

Unfortunately, as Stremlau proceeds to explain, the benefits to Bangalore are not as great as one would expect. Foreign corporations tend to isolate themselves from the local society in specially created areas such as technology parks and make no attempt to contribute to desperately needed infrastructure improvements, despite the fact that they are fast becoming the primary users and beneficiaries of such resources. These "walled communities" with little local commitment have caused upheavals in Bangalore, including riots and destruction of property owned by foreign multinationals. As a result, the government has begun to exercise caution in approving foreign-sponsored projects, and a nationalistic contingent is on the rise. Thus, it seems that a mutually beneficial trade arrangement is causing a number of problems on both sides.

Another issue that is not highlighted in Stremlau's article but has been prevalent for decades is the controversy over who does the training versus who gets the skills, that is, should Indians receiving high-level training be allowed to immigrate to places such as the United States in search of employment, or should they be forced to remain at home? This problem, commonly referred to as "brain drain," has long been a problem for many Third World countries and

TABLE 3.1 Foreign Employment as Percentage of Total Employment for 10 Largest U.S.- and Western Europe-Based Corporations

Royal Dutch	72%
Exxon	62%
IBM	48%
General Motors	36%
Nestle	97%
Ford	51%
Alcatel Alsthom	52%
General Electric	25%
Philips Electronic	88%
Mobile	44%

SOURCE: United Nations (1994, p. 6).

NOTE: Presented as cited in: Illon, Lynn. "Tracing Regional Education Trends to a Global Economy." Delivered to the Midwest and Northeastern Regional Meeting of the Comparative and International Education Society. October 28, 1995. Niagara Falls, New York.

has led many to adopt tough emigration restrictions on their elite labor force. Large multinationals frequently exacerbate this problem by luring workers through the payment of salaries (either in the country of origin or in the country of operations) that, although comparatively low by Western standards, are a lot higher than traditional opportunities available in most Third World countries. The issue becomes particularly alarming when one considers the relatively high proportion of foreign laborers employed by such multinationals. Table 3.1 provides evidence for this point.

Such companies are frequently attracted to operate in foreign countries by special provisions such as lower tax rates (in addition to the cheaper supply of labor), meaning that they make little or no contribution to the educational financing of the foreign workers they employ. Furthermore, they are able to generate higher profit margins by using lower-paid workers, frequently to the detriment of workers or job seekers in their countries of origin, meaning that the education investments made in these countries are not being fully capitalized. Once again, both governments and workers receive the short end of the stick while companies profit.

In light of situations such as these, one wonders which will prevail—globalized securitization, global free trade, or nationalistic sentiment? Given that it is more than likely this type of controversy will continue to escalate, there is a pressing need to find ways to mediate this new form of human commerce in human knowledge and technology capital. Such an effort could begin by making humans a commodity formally regulated by the General Agreement on Trade and Tariffs (GATT). Many would probably oppose this on the grounds that it evokes despicable periods of human history during which humans were traded as commodities—slavery. But palatable or not, we are once again engaging in human trade and, to use the Indian colloquialism, this body shopping is likely to continue unfettered unless some parameters are established. Given international pressures for decentralization and prioritizing education, the regulation of its "products" is unlikely. We feel it would be better to admit the situation and arrive at formally agreed-on terms of exchange than to keep burying our heads in the sand and adopting isolationist immigration and trade policies. In addition, education and social/ infrastructure investment requirements should also be entered into the GATT agreements. Such requirements should be considered a normal cost of doing business but, unfortunately, it seems that, without a regulatory mandate, most companies still prefer to exploit the resources of the foreign countries in which they "locate" but do not "live." It is time for corporations to become socially and morally responsible to *more* citizens than those who reside within their original national boundaries. Without such restrictions, we may be doomed to repeat the mistakes of the colonialist era that we now pride ourselves in having overcome.

Conclusion

This chapter has examined the linkages between education, technology, and the economy and the effect these linkages have on the present global socioeconomic order. Our focus has been to paint a broad picture in an attempt to gain a more enlightened understanding of the issues at hand. We have also attempted to raise a series of questions that could serve to guide future thinking on this subject. In summary, we hope to have been able to illustrate the following points: (a) education, technology, and the economy are inexorably and complexly linked; (b) the unprecedented speed in communications is transforming economic and education systems around the globe; and (c) unless we begin to pay close attention to these transformations and attempt to foresee their social consequences, we may be faced with a barbaric world in which inequality, the dominance of financial "mercenaries," and an uncivilized culture of impunity are the norm. We suggest that the field of education finance should engage in a more comprehensive inquiry.

Notes

1. There are currently four TIGER countries: Hong Kong, Singapore, South Korea, and Taiwan. However, given their high growth rates, Indonesia, Malaysia, and Thailand are also typically thought of as TIGER countries as well, despite the fact that their GNPs are much lower than the four TIGERs.

2. For further information see *The Economist*, 1997; any issue of the *Journal of Education Finance*; any issue of the *Comparative Education Review*; Picus & Wattenbarger, 1996.

3. Currently, there are 10 Big Emerging Markets (BEMs): the Chinese Economic Area (CEA, including China, Hong Kong, and Taiwan), India, South Korea, Mexico, Brazil, Argentina, South Africa, Poland, Turkey, and the Association of Southeast Asian Nations (ASEAN, including Indonesia, Brunei, Malaysia, Singapore, Thailand, the Philippines, and Vietnam).

4. Bill HR 2202, Section 806 (H1-B nonimmigrant) introduced by Chairman Lamar Smith and passed by the House Judiciary Committee on July 20, 1995. (For further information see: http://users.ccnet.com/ez/welcome.html)

References

Big Emerging Markets (BEMs) Home Page. (1995, July 24). About the big emerging markets? BEMs messages from Commerce Secretary Ronald H. Brown and Under Secretary Jeffrey E. Garten. (http://www.stat-usa.gov/itabems.html)

Broad, R., & Cavanagh, J. (1995-1996, Winter). Don't neglect the impoverished South. *Foreign Policy*, PA: 18-35.

Edmunds, J. C. (1996, Fall). Securities: The new world health machine. *Foreign Policy*, pp. 118-133.

Ellul, J. (1964). *The technological society*. New York: Vintage.

Enterprise computing in the twenty-first century. (1997, January). Digital Equipment Corporation Home Page, white paper. (http://www.digital.com/info/alphaserver/solutions/mainframe/datapro_wp.html)

Phillips, K. (1994). *Arrogant capital*. Boston: Little, Brown.

Picus, L. O., & Wattenbarger, J. L. (Eds.). (1996). *Where does the money go? Resource allocation in elementary and secondary schools*. Thousand Oaks, CA: Corwin.

Reich, R. (1992). *The work of nations*. New York: Vintage.

Results of 1995 search on globalization from America Online's *Compton's Encyclopedia* (key words: the economic system, the end of empire).

Rifkin, J. (1995). *The end of work*. New York: Putnam.

Smelser, N. J., & Swedberg, R. (Eds.). *The handbook of economic sociology*. Princeton, NJ: Princeton University Press.

Stremlau, J. (1996, Spring). Dateline Bangalore: Third world technopolis. *Foreign Policy*, pp. 152-168.

Texas Instruments. (1995, February). Texas Instruments: Enabling the information superhighway—today." *Texas Instruments Home Page*, news releases. (http://www.ti.com/corp/docs/pressrel/1995/95012.htm)

Thurow, L. (1992). *Head to head: The coming economic battle among Japan, Europe, and America*. New York: William Morrow.

Wall Street is fueling the consumer credit binge. (1997, March 31). *U.S. News and World Report*, U.S. News Online. (http://www.usnews.com/usnews/issue/970331/31DEBT.HTM)

Webster's Ninth New Collegiate Dictionary. (1989). Springfield, MA: Merriam-Webster.

White House Republicans reach deal on averting shutdown. (1996, January 25). *U.S. News and World Report*, U.S. News Online. News from AP: Budget Shutdown. (http://www.usnews.com/usnews/wash/0126A.HTM)

Wiens, J. R. (1995, March 9-12). *Educational reform and the globalization of the economy: "A work in progress."* Presented at the American Education Finance Association Annual Meeting, Savannah, GA.

World Education League: Who's top?" (1997, March 29-April 4). *The Economist*, pp. 21-23.

World Development Report 1995: Workers in an integrating world. (1995). Washington, DC. The International Bank for Reconstruction and Development, The World Bank. (pp. 164-165).

National and State Issues

FOUR

Common "Mythstakes" in Technology Planning

KAREN FULLERTON

The incorporation of instructional and telecommunication technologies in schools requires a strategic plan that envisions technology as integral to curriculum development and community involvement. Although technology pervades most aspects of modern life, it is underutilized in schools and needs to become better integrated with the curriculum if students are to be adequately prepared to enter the information age workforce. McKinsey & Company (1995) predict that by the year 2000 as many as 60% of American jobs may require computer literacy and networking skills. These skills include the ability to store, retrieve, and manipulate information on networks (Kaye, 1996). Unfortunately, the majority of technology plans do not establish the links between the need for technology and identifiable instructional priorities (Moersch, 1995).

For curriculum issues, teachers are the key to effective incorporation of technology in instruction. The great promise of technology is that it can improve student achievement, motivation, critical thinking, and cooperation. These accomplishments, however, depend on academic and social contexts. Research in instructional technology has identified factors that influence outcomes, including subject matter, purpose, specific hardware and software, the ratio of students to media, the amount of time available, the physical location of technology, and the aptitude and learning style of the student (Barron & Orwig, 1995; Schofield, 1995; Thompson, Simonson, & Hargrave, 1992). Teachers, therefore, not only need to improve their own technology literacy levels but also need to learn how to adapt their classroom teaching styles and extend their instructional strategies to include greater use of technologies, especially computers.

Emerging instructional technologies and telecommunications can also be harnessed to strengthen ties between schools and communities. As Lewicki (1994) describes, "Every school site is rooted in the soil of the neighborhood" (p. 84). Although schools use multiple ways to communicate and distribute information to the community, Holte (1995) calls for schools to be more proactive about using technologies to effectively connect schools and communities. Examples of these technologies include (a) e-mail to allow communication between parents and teachers, between community members and the school board, between teachers and their colleagues, and between students and resource people around the world; (b) school hotlines to provide information about lunch menus, homework assignments, closings, and events; (c) conferencing systems to link guest speakers with students in classrooms; and (d) on-line catalog systems, reference works, and computer files to link communities to each other for distant learning (Holte, 1995; McKinsey & Co., 1995).

Technology, however, is not an end in itself. The high costs of technology require a justification for technology planning that is more visionary and comprehensive than just planning hardware purchases. A successful technology plan must be a component of a larger strategy that focuses on improving learning and communication. This strategy to improve student achievement and community involvement necessitates the continuous allocation of resources to equipment and software upgrades; a commitment to ongoing training for teachers, parents, and community members; and a desire to provide local schools with access to global resources.

Unfortunately, because technology is complex and changes rapidly, the technology planning process can be misguided by myths fostered by a lack of knowledge, vendor hyperbole, faculty resistance, or pressure from the business community. Twelve common myths are addressed below that can be grouped under three main areas: technology planning, professional development, and utilization. Misinformation can result in a poor return on investments in technology. Understanding that these statements *are* myths, however, can help inform technology decision making in schools.

Myths About Technology Planning

The process of technology planning requires an investment of time and resources from numerous stakeholders, takes place at both school and district levels, emphasizes teacher training, builds in flexibility, and budgets yearly for technology purchases (Roblyer, Edwards, & Havriluk, 1997). In the long run, this planning saves time and money, helps achieve goals, builds motivation, and increases the likelihood of benefits from technology's potential to improve teaching and learning (Roblyer et al., 1997).

Some schools achieve their goals because they have technology plans that provide the flexibility needed to respond to changing needs and equipment. Others have technology plans that are never used, that never get finished being written, or that are so restrictive that they defeat their purpose. The following myths contribute to their failure.

1. *Technology plans are useless because technology changes so rapidly.*

Many plans fail because of a lack of strategic planning. Instead, they attempt to be too specific about hardware and fail to consider in any depth how technology can benefit teaching and learning. Brody (1995) describes the difference:

Most important is the *strategic plan* that can serve as the technology program's underpinning for several years. These plans are broad and conceptual in nature, recommending general directions to be taken. In contrast, *project planning*, while often complex, concentrates on the tactical details associated with actual implementation. Whereas the strategic plan is more concerned with ideas, concepts, and general directions which impact many projects, the project plan places increased attention on the specific and detailed tactical actions associated with a single project or program. (p. 7)

To avoid developing ineffective technology plans, refer to Brody (1995), Dede (1989), Hunter (1995), Lumley and Bailey (1993), Maryland State Department of Education (1991), Picciano (1994), and Solomon (1995) for useful strategies, guidelines, and models for developing and funding technology plans.

2. The computer staff should design the technology plan.

Although technology holds great promise for impacting the educational system, its nature is complicated. Increased specialization among professionals forces them to cooperate to plan under complex conditions. It is essential, therefore, that decision makers rely on input from teachers, librarians, administrators, support personnel, curriculum designers, instructional technologists, and community representatives as well as from computer specialists (Beltrametti, 1993; Brody, 1995; Roblyer et al., 1997). The degree of complication necessitates a clear vision based not on myths but on facts provided by all those involved.

3. We have to buy what businesses are using.

Given the rate at which technology changes, purchases made today are not what students will be using on the job in 1 year, 5 years, or 10 years. In the technology plan, the question that should be answered is not, "What platform and software do we purchase?" but rather, "How will technology be used and with what impact?" To date, more educational programs have been developed for use on the Macintosh, whereas more business programs have been developed for Windows. Although most software developers eventually design programs for both platforms, if the priority is for students to learn more effectively subject-specific content via computers, then software should be chosen that accomplishes that goal and equipment purchased that runs that software.

The longevity of a particular type of software also needs to be considered. It is more important that students understand what word processing software can do, what it is used for, and how it can save time and money than it is that they know how to use a specific version of a particular word processing program that will change every 6 months.

Another consideration is the amount and type of technical support that is provided with equipment warranties. More expensive, high-quality equipment that includes competent technical support and fast on-site repairs will save time and money over less expensive equipment that lacks this level of support.

4. Instructional technology means computers.

As schools develop their plans for technology integration, they need to consider how to incorporate the many varied forms of instructional and telecommunication technologies. Cuban (1986) describes instructional technology as "any device available to teachers for use in instructing students in a more efficient and stimulating manner than the sole use of the teacher's voice" (p. 4). Seels and Richey (1994) go beyond this idea of devices to define instructional technology as "the theory and practice of the design, development, utilization, management and evaluation of processes and resources for learning" (p. 1). Technologies that are available for classroom use include traditional media such as audiotapes, bulletin boards, posters, photographs, radio, slides, speaker phones, television, laser discs, and videotapes. Technologies gaining greater acceptance include CD-ROMs; distance education systems; integrated learning systems; Internet web pages; and computer-based tutorials, drills, and simulations. Productivity programs include word processing, spreadsheets, databases, e-mail, and graphics. At the administrative level, computer programs can facilitate scheduling, budgeting, accounting, inventory, and grading (Ely, 1992; Kearsley, 1990).

Schools can also communicate and distribute information to the community using different types of telecommunication technologies such as what the Maryland State Department of Education (1991) recommends: (a) for voice: two-way communication and messaging systems such as telephones, public address systems, and voice mail;

(b) for video and audio: production and transmission capabilities from the school and external providers to the school and community; and (c) for data: interconnection throughout the school with a LAN (local area network) plus two-way access through WANs (wide area networks) to resources such as bulletin boards, databases, other schools, and homes. Rhodes (1995), however, cautions that to make good use of these kinds of technologies the focus must be on the *information* that technology enables students, teachers, and administrators to access and on the *connections* that will support that information's flow.

> 5. *If we invest a lot of money in a computer lab or distance learning classroom, it will be obsolete in a year.*

What does obsolescence mean? Just because a new model arrives on the market weeks after an equipment order is placed does not mean that a purchase cannot meet learning needs and accomplish desired objectives for years to come.

Educational needs should dictate the extent to which technology is replaced. In their effort to prove they are state-of-the-art, many schools don't consider how they can repurpose older equipment, especially computers. Even if they can't access e-mail or the Internet, older computers could still be used for word processing, tutorials, or keyboard practice. For example, in a computer lab at a state university Macintosh SEs and dot matrix printers were kept in use after several PowerMac 7200s and a laser printer were installed. During a journalism class, the entire lab was used by students for word processing and printing. By integrating the use of both old and new computers, the professor was able to accommodate all the students in the class.

> 6. *Once we complete our technology plan/build our computer lab/install the distance learning classroom, we can reallocate those funds.*

A good strategic plan will consider that technology use may increase over time as users become more comfortable with and knowledgeable about technology. It will also recognize that the ongoing use of technology necessitates upgrading hardware, software, and skills; planning for repairs and maintenance; and investing in telecommunications access. This is a recognition of and commitment to the requirements of the information age. To ensure funding for long-term maintenance and growth, the strategic plan must integrate community and business needs with curriculum goals.

How can the community be persuaded to make a long-term commitment to technology in schools? One way is to no longer consider it to be restricted school property. Instead, it could be considered to be community property and made available to the community outside of regular school hours. Many schools offer low-cost computer training to community members. Some open their labs at night for students to complete schoolwork, for parents to read and send e-mail from and to teachers, and for community members to access the Internet for information or to improve their computer skills. This level of public access not only provides a valuable service but also increases the school's visibility in a positive way.

Myths About Professional Development

Teachers are the key to effective and efficient technology utilization. When technology is available, however, it is frequently used with styles of teaching that fail to maximize its full potential (National Governors' Association, 1996). This could be the result of inability, improper training, technophobia, or a lack of practice using alternative teaching strategies. Therefore, adequate professional development is needed if technology is to help schools improve learning.

7. Teachers don't take advantage of training opportunities.

Due to a lack of exposure or to prior bad experiences, some teachers are technophobic, afraid to try to use technology, especially computers, with their students. Because fear and inertia contribute to a resistance to change, the people involved must want to learn (Schofield, 1995). To help overcome resistance, professional development efforts can start by concentrating on those areas that make someone's job easier, more efficient, or more effective and should address the needs and concerns of end users. For example, a workshop for teachers on spreadsheets could begin by creating a gradebook template and emphasizing how this can save time. Then, the workshop could demonstrate different ways that students can use spreadsheets across content areas.

Administrators also need to carefully coordinate the timing of professional development activities. Too frequently, there is a long time lag between training and the opportunity to use what was learned. Equipment may be installed weeks after training is conducted or vice versa. Even those who initially are motivated eventually become discouraged by the delay.

Furthermore, training needs vary among teachers according to their skill and comfort levels. Realistic amounts of training need to be provided to achieve the desired results. According to the U.S. Congress Office of Technology Assessment (cited in McKinsey & Co., 1995), 30 hours of training are needed to successfully use technology at a basic level. For a teacher to have good operational knowledge of hardware and perform basic troubleshooting requires 45+ hours of training and 3 months of experience. For a teacher to actively develop entirely new learning techniques that utilize technology requires 80+ hours of training and 4 to 5 years of experience.

8. Teachers are not using the technology that is available.

Sometimes, instructional technology is purchased in response to a particular interest or need. A small group of teachers may be using the technology frequently, but administrators are not aware of the extent to which it is being used or think it should receive widespread use. The other teachers, however, may perceive that its use is restricted. This is exacerbated by housing the equipment in a limited-access area such as a science lab. Small-scale innovation attempts like these frequently fail because they lack a critical mass of people, funding, or equipment and the traditional system of rewards does not encourage adoption (Dede, 1989). Few people initially adopt any innovation. The rate at which others adopt innovations depends on multiple factors, including their perception of its relative advantage and complexity, their level of involvement in making decisions, and the nature of their social system (Rogers, 1983). Professional development efforts should consider these factors.

Many other barriers inhibit the adoption of technology in class-rooms. For teachers, the disadvantages include difficulty finding appropriate media or software, a lack of time for review and prepara-tion, unfamiliarity with hardware or software, inaccessibility of equipment, poor technical support, and the inability to quickly solve problems (Cuban, 1986; Evans-Andris, 1996; Schofield, 1995). Addi-tionally, equipment failures in front of students are embarrassing and a threat to teachers' sense of competence and authority (Schofield, 1995).

Schools can overcome these barriers by providing professional development opportunities, by creating incentives to learn and use technology, and by recognizing that comfort and expertise come only with experience. More important, effective leadership is needed to help turn laggards into technology adopters. Principals who dem-onstrate enthusiasm and concern for change communicate their ex-pectations, advise and support teachers, and monitor their progress provide a source of motivation for teachers (Evans-Andris, 1996; Moersch, 1995).

9. *Teachers will change from "knowledge experts" to "knowledge facilitators" working with students who follow individual and team learning plans.*

In this myth, students are self-directed in their search for knowl-edge—which assumes an unqualified love of knowledge and a great deal of self-motivation. Loveless (1996) points out that students rarely exhibit a love of learning and must be motivated and guided by their teachers.

This myth also presumes that administrators, teachers, and stu-dents can readily accept different classroom and school cultures. Typically, teachers manage about 30 students in the classroom. They can accomplish this because they are viewed by students as both information experts and authority figures. Changing their role to knowledge facilitators does not diminish those roles. Instead, it adds the dimension of managing the labor of students who may be working at different speeds on different tasks. The culture of these classrooms can also be more active, energetic, and noisy and may require greater flexibility in scheduling and assessment methods. Professional development efforts must address these concerns.

Myths About Technology Utilization

The desired outcome of technology planning is to impact learning in positive ways by investing in appropriate technology and effective professional development. One determination of whether technology is used successfully or not depends on how equitably it provides access to information to all students. Access to technology, however, is not where it goes but how it is used (National Governors' Association, 1996). The ultimate determinant of success is not reaching some ratio of equipment to students but ensuring that knowledgeable teachers are using technology efficiently and effectively to meet lesson goals and objectives.

> 10. *Even though we don't have a technology plan, we need to get at least one or two computers in every classroom.*

Getting computers into schools and employing them in instruction are not the same thing (Loveless, 1996). A computer in every classroom allows schools to show off on parents' night, but without adequate professional development, Internet connectivity, or technical support staff these computers are underutilized. Although some teachers may use them for word processing or record keeping, for instructional benefits to be realized clearly defined objectives must be established for how computers will be used and integrated with the curriculum.

Teachers in general are not very knowledgeable about how to integrate any type of technology into their lessons, much less computers or software. When Schofield (1995) studied utilization in classrooms where computers were available for student use, she found that teachers are not easily able to solve the organizational and scheduling problems posed by a low ratio of computers to students. For example, when some students are allowed to use the computers in an adjacent classroom, they aren't able to ask the teacher questions or get assistance solving technical problems. Also, printers are noisy and distracting to those students who are trying to accomplish other tasks.

Another problem is that it is ineffective to attempt to teach a class with one computer without a data display for large screen projection. Small monitors are useless for group instruction, yet few schools invest in projection technology so the entire class can see what the teacher is demonstrating. It is possible for groups of students to work on different tasks simultaneously and rotate computer usage among groups. But as Loveless (1996) puts it, "Coming up with imaginative, educationally inclusive ways for groups of students to use a class-room's few computers remains a difficult task" (p. 30).

11. Teachers will be replaced by computers.

This myth ignores the culture of the classroom and doesn't consider how teachers meet important social and emotional needs for their students. Its persistence as a myth seems to stem from efforts in artificial intelligence that are applied to instruction. Although highly advanced computer-based tutorials that utilize artificial intelligence technology are being designed that adapt to students' individualized learning styles and needs, they are expensive and labor intensive to design. It will be many years before this technology is common in classrooms and available for all grade levels and curriculum areas. It is also doubtful they will be able to respond the way teachers do to changes in students' readiness for learning and the social context of the classroom. As Schofield (1995) cautions,

> Computers . . . are social as well as technological objects, and their use is subject to the vagaries of the social milieu in which they are available for use, although over time they may profoundly influence that milieu. If they are too out of step with existing practice, they are likely not to be used extensively, or they will be used in ways that do not take full advantage of their potential. (p. 228)

Understandably, teachers are concerned about the role of technology in their classrooms. But it is unlikely that these computer-based systems will replace them. Teachers best understand their students, the classroom setting, and the complexity of the school environment and design instructional tasks accordingly.

12. *Technology will transform education.*

Teachers, students, and communities can transform education—technology merely provides some of the tools and processes that can support educational reform and create pressure for change. Years of media research have shown that technology and media in the classroom do not benefit all students equally nor result in across-the-board increases in achievement. Technology in schools, however, can help teachers become more efficient and effective and can help improve student motivation and engagement with the learning process. Additional research in the interactions between technology and learner abilities will inform teachers of the best ways to use technology to benefit students.

Conclusion

Purchasing technology is not an end in itself. To plan for integrating technology into a school's curriculum requires a strategic vision for that curriculum. That vision becomes the foundation for a plan that demonstrates how technology supports instruction and comprehensively addresses the issues of incorporating many different kinds of instructional technology in classroom instruction; of providing suitable professional development that meets the needs of different teachers; of purchasing high-quality software and media for all disciplines; and of providing links between schools, parents, and communities for the long-term benefit of all.

For technology to be worth its cost, the technology plan also has to show how it helps to meet the needs of the community by preparing students for work in the information age. If it continues to be considered as something separate, Brody (1995) cautions that "students will have greater difficulty in accepting and understanding one of the principal tenets of modern technology, the fact that technology is part of virtually all aspects of modern society" (p. 45).

Dealing with technology is not easy. But no one can ignore it any longer or deny that it is a task that needs to be done. Strong leadership is required to deal with competing demands for scarce budgets, to motivate teachers to incorporate technology into their teaching strategies, and to make creative use of emerging technologies.

References

Barron, A. E., & Orwig, G. W. (1995). *New technologies for education: A beginner's guide* (2nd ed.). Englewood, CO: Libraries Unlimited.

Beltrametti, M. (1993). Human infrastructures to enable the magic of information technology. In T. A. Brubaker (Ed.), *The magic of technology: Proceedings of the National Educational Computing Conference* (pp. 11-15). Orlando, FL.

Brody, P. J. (1995). *Technology planning and management handbook: A guide for school district educational technology leaders.* Englewood Cliffs, NJ: Educational Technology Publications.

Cuban, L. (1986). *Teachers and machines: The classroom use of technology since 1920.* New York: Teachers College Press.

Dede, C. (1989). Planning guidelines for emerging instructional technologies. *Educational Technology, 29*(4), 7-12.

Ely, D. P. (1993). Trends and issues in educational technology. In M. Shaw & E. Roger (Eds.), *Aspects of education and training technology: Vol 26. Quality in education and training* (pp. 237-240). London: Kogan Page.

Evans-Andris, M. (1996). *An apple for the teacher: Computers and work in elementary schools.* Thousand Oaks, CA: Corwin.

Holte, J. (1995). Electronically connecting your community and schools: Why should you? *Learning and Leading With Technology, 22*(8), 43-46.

Hunter, B. M. (1995). *From here to technology: How to fund hardware, software, and more.* Arlington, VA: American Association of School Administrators. (ERIC Document Reproduction Service No. ED 385 000)

Kaye, J. C. (1996, April). *Characteristics of effective networking environments.* Paper presented at the annual meeting of the American Educational Research Association, New York. (ERIC Document Reproduction Service No. ED 394 502)

Kearsley, G. (1990). *Computers for educational administrators: Leadership in the information age.* Norwood, NJ: Ablex.

Lewicki, J. A. (1994). Natural metaphors of change for sustainable rural school communities. In G. P. Karim & N. J. Weate (Eds.), *Toward the 21st century: A rural education anthology* (Vol. 1, pp. 83-85). Rural School Development Outreach Project. Washington, DC: Office of Educational Research and Improvement. (ERIC Document Reproduction Service No. ED 401 073)

Loveless, T. (1996, February). *Why aren't computers used more in schools?* (Faculty Research Working Paper Series, R96-3). Cambridge, MA: Harvard University. (ERIC Document Reproduction Service No. ED 392 131)

Lumley, D., & Bailey, G. D. (1993). *Planning for technology: A guidebook for school administrators.* New York: Scholastic.

Maryland State Department of Education. (1991, March). *Model educational specifications for technology in schools.* Baltimore: Author. (ERIC Document Reproduction Service No. ED 354 864)

McKinsey & Co. (1995). *Connecting K-12 schools to the information superhighway*. Washington, DC: Author. (ERIC Document Reproduction Service No. ED 393 397)

Moersch, C. (1995). Levels of technology implementation (LoTi): A framework for measuring classroom technology use. *Learning and Leading With Technology, 23*(3), 40-41.

National Governors' Association. (1996, March). *Technology and education standards: Issue brief*. Washington, DC: Author. (ERIC Document Reproduction Service No. ED 394 518)

Picciano, A. G. (1994). *Computers in the schools: A guide to planning and administration*. New York: Merrill

Rhodes, L. A. (1995). Technology-driven systemic change. *Learning and Leading With Technology, 23*(3), 35-37.

Roblyer, M. D., Edwards, J., & Havriluk, M. A. (1997). *Integrating educational technology into teaching*. Upper Saddle River, NJ: Prentice Hall.

Rogers, E. M. (1983). *Diffusion of innovations* (3rd ed.). New York: Free Press.

Schofield, J. W. (1995). *Computers and classroom culture*. Cambridge, UK: Cambridge University Press.

Seels, B. B., & Richey, R. C. (1994). *Instructional technology: The definition and domains of the field*. Washington, DC: Association for Educational Communications and Technology.

Solomon, G. (1995). Planning for technology. *Learning and Leading With Technology, 23*(1), 66-67.

Thompson, A. D., Simonson, M. R., & Hargrave, C. P. (1992). *Educational technology: A review of the research*. Washington, DC: Association for Educational Communications and Technology.

FIVE

The Coming Crisis in Student
Access to Educational Technology
REVISIONING THE STATE AND
FEDERAL ROLES IN FUNDING

FAITH E. CRAMPTON

The United States is approaching an educational crisis that threatens the social and economic well-being of the nation. As an increasing number of jobs demand technology skills, student access to educational technology in elementary and secondary schools becomes essential. Although in 1984 only 25% of jobs required technology skills, in 1993 the percentage had almost doubled to 47%, and for the year 2000 it is estimated that more than 60% of jobs in the nation will require these skills (The Children's Partnership, 1994). At the same time, tremendous inequities in student access to technology exist within and across states. To delay action is to risk further socioeconomic stratification in the country, at worst, the emergence of a bilevel society

AUTHOR'S NOTE: The views expressed by the author in this chapter are not necessarily those of the National Education Association.

77

of technology haves and have-nots, where those with technology skills compete for the well-paying jobs while those without them face subsistence employment, creating a permanent underclass.[1]

To address this critical policy issue, this chapter proposes a re-visioning of federal and state roles in funding adequate and equitable student access to educational technology. This revisioning involves forging a new federal-state partnership with the federal government taking a leadership role in setting broad goals for student access to technology and serving as senior partner in funding. The chapter is divided in five parts: The first describes the current state and federal roles in providing student access to technology. The second section estimates the fiscal resources needed. The third section analyzes the current policy position of state responsibility for student access to educational technology and posits a revisioning of state and federal roles. The fourth section proposes a revenue source to provide a portion of funding. The final section presents conclusions and policy recommendations.

Current Federal and State Roles

In February 1996, the Technology Literacy Challenge was an-nounced by President Clinton and Vice President Gore, setting forth four national goals: (a) all teachers in the nation will have the training and support they need to help students learn to use computers and the information highway; (b) all teachers and students will have modern multimedia computers; (c) every classroom will be connected to the information highway; and (d) effective software and on-line learning resources will be an integral part of every school's curriculum (U.S. Department of Education, 1996). These admittedly ambitious goals, to be achieved by the year 2000, carry cost estimates that range from $50 to $100 billion dollars (U.S. Department of Education, 1996). A maximum of $10 billion dollars has been pledged, not appropriated, at the federal level, ostensibly leaving the remainder of funding to the state and local school districts.

Yet, states vary substantially in planning and funding activities for student access to educational technology. Table 5.1 presents a state-by-state summary of statewide planning and funding of educational

TABLE 5.1 State Planning and Funding of Technology

State	State Plan	Highlights	State Funding	Notes
Alabama	N	n.a.	n.a.	Governor's commission developing plan
Alaska	N	n.a.	n.a.	Finalizing statewide plan
Arizona	Y (1991)	Rural school connectivity initiative	$6.8 million (pending)	Plan under consideration by legislature
Arkansas	N	n.a.	n.a.	
California	Y	Regional allocation of funds	$100 million (proposed, 1997)	
Colorado	Y (1995)	Emphasis on staff development	n.a.	
Connecticut	Y (1995)		$10.4 million (competitive grants, 1995)	
Delaware	Y	Linkage of every classroom goal	Goals 2000 Funds	
District of Columbia	Y (1991)	Creation of Center for Innovative Technology to assist schools	Title I Funds	
Florida	Y (1989)	Educational Facilities Infrastructure Improvement Act	$117 million (1995)	

(Continued)

79

TABLE 5.1 (Continued)

State	State Plan	Highlights	State Funding	Notes
Georgia	Y	Educators' Technology Training Commission	$50 million	
Hawaii	Y	Hawaii Net Day (1/11/96)	$2 million NSF Grant (1994)	
Idaho	Y (1992)	Idaho Education Technology Initiative of 1994	$10.4 million; $3 million in competitive grants (1994)	Plan not funded
Illinois	N	n.a.	$15 million (1996)	
Indiana	N	n.a.	$10,000 planning grants to qualifying school districts	State-required 5-year school district technology plans
Iowa	Y	Iowa Communications Network	$36 million (1996 and 1997)	
Kansas	N	n.a.	n.a.	State-developed technology planning guide for schools
Kentucky	Y	Recommended home connections for educational use	$20 million (1995)	
Louisiana	Y	LaNET (wide area network)	n.a.	
Maine	N	n.a.	$15 million for distance learning network (1995)	Planning efforts are under way

Maryland	Y (1995)	Statewide technology inventory	n.a.	
Massachusetts	Y (1994)	Emphasis on staff development	$60 million (proposed)	
Michigan	N	n.a.	$10.5 million (1995)	
Minnesota	N	n.a.	$5.4 million in grants (1996-1997 biennium)	Plan being developed by a task force established by the state department of education
Mississippi	Y (1995)	Technology standards for teachers	$26.8 million (1994)	
Missouri	N	n.a.	$5 million (1995)	
Montana	N	n.a.	$100,000 (1995)	Plan being developed by a technology task force appointed by the governor and state superintendent of education
Nebraska	Y (1996)	Universal school access to Internet in 1997	n.a.	
Nevada	N	n.a.	$400,000	
New Hampshire	N	n.a.	n.a.	Guidelines for local technology plans being drafted

(Continued)

TABLE 5.1 (Continued)

State	State Plan	Highlights	State Funding	Notes
New Jersey	Y (1993)	Technology specifications for school facilities	$10 million (1997); $1.3 million, competitive grants	
New Mexico	Y (1995)	Equal funding for every child	$9.50 per student (1996)	
New York	Y (1990)	Emphasis on staff development	n.a.	
North Carolina	Y (1994)	Comprehensive infrastructure goal	$42 million (1996)	
North Dakota	N	n.a.	n.a.	Planning activities continue
Ohio	Y (1992)	Focus on poor school districts	$275 million (proposed)	
Oklahoma	N	n.a.	n.a.	
Oregon	Y (1992)	n.a.		
Pennsylvania	N	n.a.	$100 million (over 3 years)	"Project Link to Learn" proposed by governor
Rhode Island	N	n.a.	n.a.	Plan complete and awaits board of regents approval
South Carolina	Y (1995)		State foundation	

South Dakota	N	n.a.	none	
Tennessee	N	n.a.	$98 million (since 1994)	
Texas	Y (1988)	Review of plan by task force (1996)	$30 per student; $150 million for infrastructure (over 10 years)	
Utah	N	n.a.	$33 million (anticipated)	State requires 5-year school district technology plans with annual updates
Vermont	Y	Distance learning program for advanced placement courses	n.a.	
Virginia	Y (1988)	Updated (1995)	$75 million (proposed)	
Washington	Y (1994)		n.a.	
West Virginia	N	n.a.	$3.9 million (1996)	
Wisconsin	N	n.a.	State foundation; $10 million grant program	
Wyoming	N	n.a.	none	State in initial stages of planning

SOURCE: Adapted from U.S. Department of Education (1996, pp. 60-68).

technology.[2] As of 1996, 27 states and the District of Columbia had completed statewide plans for educational technology. The earliest plans were developed in 1988 by Texas and Virginia; each subsequent year has seen another state or two complete plans, peaking in 1995 with six states—Colorado, Connecticut, Maryland, Mississippi, New Mexico, and South Carolina. Of the 23 that do not have plans, 10 are in some stage of development. Nine states report no planning activity, and four—Indiana, Kansas, New Hampshire, and Utah—have chosen to leave planning to local school districts. The last two categories include 13 states; that is, approximately one-fourth of the states have not given any indication that they intend to develop statewide plans. This group of states, in particular, raises concerns about adequate and equitable student access to technology.

The means of funding educational technology also vary widely among states, with a number making a yearly or biennial decision as to an appropriation, if any. It is important to remember that states differ in both their fiscal capacity and their willingness to fund educational technology. Ten states and the District of Columbia have chosen to pursue nontraditional methods in addition to or in place of regular appropriations. Of these, six states—Connecticut, Idaho, Indiana, Minnesota, New Jersey, and Wisconsin—offer competitive grant programs that rely on school district application. Two states, South Carolina and Wisconsin, have established state foundations to attract private donations. Delaware has taken advantage of Goals 2000 funds, and Hawaii has utilized a National Science Foundation grant. The District of Columbia presently uses Title 1 funds. Highlights of plans indicate that funding priorities differ also. For example, several states—Colorado, Georgia, Massachusetts, and New York—have placed emphasis on staff development to enable teachers to integrate technology into the classroom. Arizona emphasizes rural school connectivity, and Kentucky has given home connections a high priority.[3] Nebraska has set the goal of connecting every school, and Delaware, every classroom. North Carolina is focusing on providing infrastructure for technology whereas Vermont's focus is distance learning.

Although these states may be applauded for their tenacity in searching out creative funding mechanisms, such approaches leave students within and across states with unequal access to technology. A simple but potent comparison captures the magnitude of the inequity: The results of a national survey found student/computer ratios, also termed "computer density," ranged from a low of 7 students per computer in Wyoming to a high of approximately 25 in Oklahoma (U.S. General Accounting Office, 1995). At the same time, the U.S. Department of Education recommends one computer for every five students as a baseline to ensure that students have sufficient access to use computers on a regular basis during the school day (U.S. Department of Education, 1996).

Estimating the Fiscal Resources to Provide Student Access to Educational Technology

Estimating the level of fiscal resources to provide adequate and equitable student access to educational technology is a formidable task. Two recent studies using different methodologies have attempted to do so. The first, developed by McKinsey & Company, Inc. in 1995 for the U.S. Advisory Council on the National Information Infrastructure of the National Telecommunications and Information Administration of the U.S. Department of Commerce, was incorporated in the department's report, *Kickstart Initiative: Connecting America's Communities to the Information Superhighway*; the second study was published in 1996 by the RAND Corporation (Glennan & Melmed, 1996) for the Office of Science and Technology Policy and the Office of Technology of the U.S. Department of Education. Although the methodologies of both studies are defensible from a research point of view, the McKinsey methodology is more helpful from a policy perspective for several reasons: (a) it presents policymakers with an average school prototype against which to benchmark other schools and technology costs; (b) it is more specific as to what new dollars will buy; (c) it gives policymakers a range of models for utilization of technology from which to choose, with each model clearly differentiated; (d) it distinguishes clearly between initial investment versus ongoing operating costs in projections; and (e) it recognizes the importance of investing in professional development on a continuous basis to maximize the effectiveness of educational technology.

The notion of the average school is intuitively appealing as well as practical from a policy perspective. The average school prototype serves as a benchmark, for example, allowing a legislator or governor to compare and contrast the prototype to schools in her or his state. In contrast, the RAND study utilized a small group of eight schools designated as "technology rich" and extrapolated a national cost for educational technology. The eight exemplary schools singled out by RAND were quite different along a number of dimensions, from size and demographics to the density and uses of technology. With regard to computer density, the single largest initial investment in technology, ratios ranged from one computer for every 2 students to one for 11 students.

Second, the RAND study lacked the specificity of the McKinsey study with regard to what funding would buy. Whereas RAND used cost categories for projection based on accounting data from the technology-rich schools to arrive at percentage increases in overall funding for access to technology, the McKinsey study specified six cost components: (a) connection to school, (b) connection within school, (c) hardware, (d) content, (e) professional development, and (f) systems operation. For each model, cost components were operationalized in terms of number of computers, type of local area network, and line connection (e.g., telephone vs. T-1), and then costed.

Third, the McKinsey study presented five models from which to choose, ranging from a traditional "lab" model, where technology was centralized, to a "desktop model," with a computer on every student's desk. In each of these models, it is clear to the policymaker what funds would purchase.

Fourth, distinguishing between the initial investment versus ongoing technology costs is critical. Partitioning out initial investment costs enables policymakers to see the magnitude of funding needed to level the playing field for student access to technology. In addition, as the initial investment falls largely within the category of capital outlay, it could be funded through bonded indebtedness. On the other hand, ongoing costs of technology with regard to maintenance, supplies, personnel, and professional development are considered operating costs that require annual appropriations.

Finally, the McKinsey model emphasizes the importance of professional development through its recognition of levels of expertise developed by teachers, ranging from entry-level skills to creative, inventive uses of technology that are acquired over years of learning and use. This framework is supported through the structure of funding for technology that recognizes that although the cost of hardware is the largest initial cost, professional development is the largest ongoing cost.

The central questions then become which of the five models presented in the McKinsey study best ensures adequate and equitable student access to educational technology and how much does it cost? The five models are (a) lab, (b) lab plus, (c) partial classroom, (d) classroom, and (e) desktop. See Table 5.2 for a description of each model. The authors believe that a substantial number of schools already meet or exceed the traditional computer lab model although some of their computers may be antiquated, for example, lacking multimedia and connectivity capabilities. At the other extreme, the desktop model, which sets a student to computer ratio of one is prohibitively expensive, with an initial investment of $165 billion, more than three times that required for the classroom model. See Table 5.3 for initial investment and ongoing costs of each model. A step above the lab model is the lab plus model, which adds a computer and modem for each teacher. This model does not give students additional access to technology. The fourth model is termed a partial classroom model, with half of a school's classrooms containing computers in a five to one ratio with students. This model forces an unpleasant policy decision about which classes and teachers have access to technology. For students to have regular, daily access to computer technology, the classroom model, with a ratio of five students per computer in every classroom, with all computers networked and connected to the information highway, allows meaningful student access.

The classroom model requires an initial investment of $47 billion, which the authors recommend be phased in over 10 years. If equipment and infrastructure were financed through bonded indebtedness, allowing a multiyear repayment period, a 10-year period may be reasonable for components such as wiring, furniture, and remodeling. Five years more accurately reflects the useful life of computers and peripherals, however. Although McKinsey & Company (1995; U.S. Department of Commerce, 1995) project annual operation and maintenance would be prorated based on phasing in equipment over a 10-year period, peaking in the 10th year at $14 billion, 5 years is a more realistic scenario, with $14 billion representing an annual cost, as most

TABLE 5.2 Models for Student Access to Technology

Lab Model	Lab Plus Model	Partial Classroom Model	Classroom Model	Desktop Model
• Single room	• Single room	• Half of class-rooms have 1 computer per 5 students	• All classrooms have 1 com-puter per 5 students	• All classrooms have 1 com-uter per stu-dent
• 25 computers	• 25 computers	• Ethernet LAN across and within all classrooms	• Ethernet LAN across and within all classrooms	• Ethernet LAN across and within all classrooms
• Ethernet LAN • 10 telephone lines	• Ethernet LAN • 10 telephone lines • 1 computer and modem per teacher	• T-1 connection	• T-1 connection	• T-1 connection

SOURCE: U.S. Department of Commerce (1995, p. 93).

TABLE 5.3 Cost Estimates for Models for Student Access to Technology (in billions of dollars)

	Lab	Lab Plus	Partial Classroom	Classroom	Desktop
Initial Investment	11	22	29	47	165
Ongoing Cost	4	7	8	14	35

SOURCE: U.S. Department of Commerce (1995, p. 93).

of the technology would be installed in the first year. This translates into a cost of $117 billion over a 5-year period.[4]

These estimates take into consideration technology needs due to projections of increased enrollment over the next decade. According to the U.S. Department of Education, public elementary and second-ary school enrollments rose 8.6% between 1987 and 1993 (U.S. Depart-ment of Education, 1995). Between 1993 and 2005, it is estimated enrollments will increase another 14.5% (U.S. Department of Educa-tion, 1995). The projections also give some consideration to the cost of making aging schools technology ready.

Revisioning the State and Federal Roles in Student Access to Educational Technology

This section contrasts two policy options to address student access to educational technology, analyzing the potential of each to achieve the goal of adequate and equitable student access. The first policy option represents maintenance of the status quo of state responsibility; the second proposes a revisioning of a new federal-state partnership, with the federal government as senior partner, based on framing student access to educational technology as a national crisis.

Historically, education and its funding have been considered state responsibilities, based upon the 10th Amendment to the U.S. Constitution, which states, "The powers not delegated to the United States by the Constitution, nor prohibited by it to the States, are reserved to the States respectively, or to the people." With regard to education funding, states' rights were reaffirmed in 1973 in *San Antonio Independent School District v. Rodriquez*. Here, one finds support for the status quo where decisions regarding educational technology, including funding, would remain at the state level. This position encompasses a state's prerogative to do nothing, or the state may choose to delegate responsibility to the local school district where the ability to provide student access is premised on local wealth. In this scenario, the federal government plays a minimal role at most so as not to interfere with states' rights. It is important to consider the long-term consequences of this policy option, however. Recent research indicates that children in poverty and in rural areas have less access to educational technology, such as Internet access, than those in suburban areas (U.S. Department of Education, 1997). Average computer density varies more than 300% across states (U.S. General Accounting Office, 1995). As previously stated, nine states give no indication that they plan to develop technology plans, and four more leave school districts to their own devices. The evidence is compelling that allowing student access to technology to remain a state choice will lead to widening interstate and intrastate inequities.

If continuance of the status quo is unsatisfactory, what policy basis can be drawn for greater federal involvement through a new federal-state partnership? There have been instances where Congress has deemed an educational issue so critical to the nation that it transcended states' individual rights and resulted in ground-breaking federal legislation and funding based on what is commonly referred to as the "general welfare clause" of the Constitution, Article I, Section 8. Three major examples are the National Defense Education Act (NDEA) of 1958, the Elementary and Secondary Education Act (ESEA) of 1965, and the Education of All Handicapped Act (EHA) of 1975. With the NDEA, Congress reacted to the launching of Sputnik by the former Soviet Union, concluding that the United States was losing the space race, a threat to national strength and security. To that end, the NDEA funded a nationwide campaign to upgrade the education of students in mathematics, science, and foreign languages while providing financial aid for teacher preparation. The ESEA, following on the heels of the Civil Rights Act of 1964, addressed social injustice in the provision and funding of education through targeted funding for students in poverty. Finally, the EHA remedied inequities in education of students with disabilities by requiring that students regardless of handicapping condition receive a free and appropriate education in the least restrictive environment.

Student access to educational technology represents a situation as critical as those facing the nation in 1958 in global competitiveness and in 1965 with social justice. The creation of a bifurcated socioeconomic class structure, where those with expertise in technology have access to greater employment opportunities and a higher quality of life, leaving behind those who did not have the good fortune to gain such skills in the educational system, represents a threat to our nation's economic and social well-being, as well as to the U.S. strength as a world leader. A federal-state partnership forged with targeted federal funding contingent on an ability-to-pay state match would fund student access to educational technology in order to achieve national technology literacy criteria. To maximize accountability, states would be required to submit technology plans that elaborate how they will achieve the technology literacy goals. Given the attractiveness of the federal funding, few states would choose not to participate, in the same vein as Goals 2000. But, as with Goals 2000, some states might not participate, so it is necessary to build in a provision for school districts in nonparticipating states to be included if they desire.

A Potential Source of Revenue

A new federal-state partnership will carry a substantial price tag. With the present emphasis on a balanced federal budget, it is unlikely that funding can be found entirely within current federal revenues. Hence, new revenue sources must be tapped. This section explores one potential source of revenue for funding adequate and equitable student access to educational technology—a dedicated sales tax on interstate electronic commerce, which although imposed and collected at the federal level has as its major purpose to recover and redistribute tax revenues that might otherwise be lost to states.

Industry analysts most commonly divide economic activities on the Internet into five categories: (a) Internet infastructure, (b) consumer content, (c) business content, (d) on-line trade, and (e) financial services (Forrester Research Inc., 1996). The Internet infrastructure includes the purchase and leasing of computer hardware and software as well as charges and fees for telecommunications access, Internet access, and consumer on-line services. Consumer content refers to advertising and subscriptions and fees paid by on-line providers, such as America Online or CompuServe, for exclusive rights to content of interest to consumers. Business content refers to information services now supplied on paper or through proprietary networks, such as Lexis/Nexis and Dow Jones News/Retrieval. On-line trade refers to interactive retail, including the purchase of tangible goods over the Internet, such as household goods and clothing, and intangible products, such as the downloading of software, music CDs, and videos (Rothman, 1996; Sandberg, 1996; Sherman, 1994). Included in this category are transaction charges and fees from the migration of proprietary EDI (electronic data exchange) systems to the Internet, which has already begun. These systems will move to the Internet as a means of lowering costs. The final category is financial services, which includes consumer use of banking services and management of fiscal assets, such as mutual funds.

The main rationale for taxation of the Internet economy lies with principles of sound tax policy, in particular, with maintenance of a uniform taxation system that distributes tax burden equitably (Musgrave & Musgrave, 1989). In terms of tax equity, there exists no reasonable explanation for exempting a transaction in cyberspace from taxation if it would be taxed as a nonelectronic transaction. In the nonelectronic economy, it is mainly states that tax purchases of property and services via sales taxes. Their ability to do so is limited with regard to interstate commerce, however. Given the ease with which companies may cross state lines to market products and services over the Internet, it is reasonable to assume that a substantial portion of existing and future electronic commerce will take place over state lines. The interstate nature of electronic commerce poses serious problems for states because they are limited to collecting sales tax from merchants who have a physical presence in their state, a concept called *nexus*.[5]

To place this discussion in context, it is necessary to evaluate the revenue potential of electronic commerce. One industry group, Forrester Research, has estimated the revenue potential of electronic commerce or the Internet economy to approach $90 billion dollars by the year 2000. Their projections are broken down by the categories presented earlier: (a) Internet infrastructure, $14.2 billion; (b) consumer content, $2.8 billion; (c) business content, $6.9 billion; (d) on-line trade, $21.9 billion; and (e) financial services, $46.2 billion (Interactive Services Association, 1996). As an example of growth over the next 5 years, interactive retail, a subcategory of on-line trade, was analyzed separately, revealing that it is projected to grow from $.52 billion in 1996, to $6.9 billion in 2000, an average annual growth rate of more than 100%.

It is possible to create a range of scenarios to project potential loss of sales tax revenues to states attributable to interstate interactive retail activity over the next 4 years. Two scenarios are presented below, one conservative and another worst-case. These projections are based on interpretation of reports of business activity in industry reports, professional publications, and print media. For each scenario, two estimates are calculated and then added for total potential sales tax revenue loss. The first is for the amount of sales tax revenue loss due to existing electronic merchants, and the second represents a portion of revenues generated by traditional in-state merchants who will shift to electronic commerce and expand their customer base beyond state lines.

In the conservative scenario, it is projected that lost sales tax revenues from existing electronic merchants will increase 10% annually, from 20% of the revenue base in 1997 to 50% in the year 2000, assuming a 5% sales tax.[6] The portion of revenues lost due to the migration of traditional in-state merchants to multistate electronic merchants will increase .25% annually, from .25% of sales tax revenues in 1997 to 1% in the year 2000. Although in 1997 the total loss would be only $340 million, it would rise to $1.08 billion in 1999 and $1.49 billion in the year 2000. (See Appendix 5.A for details of the calculations.)

In the worst-case scenario, it is projected that lost sales tax revenues from existing electronic merchants will increase 10% annually, from 50% of the revenue base in 1997 to 80% in the year 2000, assuming a 5% sales tax rate. The portion of revenues lost due to the migration of traditional in-state merchants to multistate electronic merchants will increase 1% annually, from 1% of sales tax revenues in 1997 to 4% in the year 2000. Although in 1997 the total loss would be only $ 1.35 billion, it would rise to $4.12 billion in 1999 and $5.56 billion in the year 2000. (See Appendix 5.B for details of the calculations.)

Because of issues of nexus, this erosion of the state sales tax base appears unavoidable. Although projections for lost revenue in 1997 seem modest, between $340 million and $1.35 billion, by the year 2000 they will have grown to between $1.49 billion and $5.56 billion. In addition, industry analysts project continued growth in electronic commerce well beyond the beginning of the 21st century. Although states cannot access this revenue, the federal government may tax interstate commerce. In this case, a federal tax could seek to recoup for states revenue lost due to a shift in the economy that is beyond their control.[7] At the same time, revenues from the tax could be dedicated to funding student access to technology and distributed based on a state-matching formula that takes into account state needs and fiscal capacity.

Conclusion and Policy Recommendations

Rather than waiting for an educational crisis to engulf the nation, policymakers may choose to view student access to technology as an opportunity to proactively seek a solution through a revisioning of the federal and state roles. At the federal level, four broad technology literacy goals already have been set in place. A new federal-state partnership where the federal government acts as a senior partner to fund educational technology through a modest tax on interstate electronic commerce with a required state match could ensure adequate and equitable student access to educational technology. Dedication of those revenues to educational technology would create a revenue stream over time to address the many facets of student access to technology—not only elements such as computers and connectivity but also staff development and support to ensure that technology is fully integrated into classrooms.

Notes

1. Although the major rationale for student access to technology presented in this chapter is limited to socioeconomic factors, the benefits are not limited to those. It is important to note that as more products, services, cultural pursuits, and entertainment are available on-line, one's quality of life will become more linked to the ability to access and use those entities. In addition, this chapter does not address a growing body of research that supports improved student achievement through effective use of technology in the classroom, particularly for low-achieving students.

2. The information in Table 5.1 was gathered by the U.S. Department of Education and the Software Publishers Association in spring 1996. Although the information is limited to an overview of statewide technology plans, it is at present the most comprehensive information available. In addition, the information on state funding is inconsistent in that it does not systematically present all of the funding for technology for each state. It is based on state responses, however.

3. *Connectivity* here refers to computer connections to the Internet and World Wide Web. Commonly referred to as the "information superhighway," the Internet began as a system connecting the U.S. government and academic institutions (U.S. Department of the Treasury, 1996). The World Wide Web represents a multimedia, hypertext application of the Internet.

4. Because of the phase-in of the initial investment over a 5- to 10-year period and the subsequent prorating of ongoing costs, McKinsey & Company do not give a cumulative total for initial investment and ongoing costs.

5. The Commerce Clause and the Due Process Clause of the U.S. Constitution are the basis of the concept of nexus. Three U.S. Supreme Court decisions define nexus and its relationship to state sales and use taxes: (a) *National Bellas Hess, Inc. v. Dep't of Revenue*, (b) *Complete Auto Transit v. Brady*, and (c) *Quill Corp. v. North Dakota*. In 1967, *National Bellas Hess* established the definition of nexus as physical presence with regard to the liability of a business to collect sales or use tax on out-of-state mail order transactions. Most recently, in 1992, *Quill Corp.* reaffirmed the portion of the *National Bellas Hess*

decision that defined nexus as physical presence. The 1976 decision in *Complete Auto Transit* outlined four criteria for deciding whether a state tax meets the constitutional requirements of the Commerce Clause, with nexus as one of the criteria. To be constitutional, the state tax must (a) apply to an activity with a substantial nexus with the taxing state, (b) be fairly apportioned, (c) not discriminate against interstate commerce, and (d) be fairly related to the services provided by the state.

6. Five percent represents the median state sales tax rate for 1995 based on U.S. Census Bureau data.

7. Although tax collection issues have been raised as an important issue ("How Does the Taxman," 1995; Mills, 1996), it is beyond the scope of this chapter to address them here. For a more thorough discussion, see Kutten (1996) and Picus and Crampton (1997).

References

The Children's Partnership. (1994, September). *America's children and the information superhighway.* Santa Monica, CA; Washington, DC: Author.

Civil Rights Act of 1964, Pub. L. No. 88-352.

Complete Auto Transit v. Brady, 430 U.S. 274 (1976).

Education of All Handicapped Act of 1975, Pub. L. No. 94-142.

Elementary and Secondary Education Act of 1965, Pub. L. No. 89-910.

Forrester Research, Inc. (1996, November). People and technology strategies: The Internet economy. *The Forrester Report.* Cambridge, MA: www.forrester.com.

Glennan, T. K., & Melmed, A. (1996). *Fostering the use of educational technology: Elements of a national strategy.* Washington, DC: RAND.

How does the taxman collect on Internet transactions? No one seems to know. (1995, February 20). *Computergram International,* CGN02200008.

Interactive Services Association. (1996, October). *Interactive marketing facts.* www.isa.net.

Kutten, L. J. (1996). Is the Internet a threat to sales and use tax collection? *State Tax Notes,* 11(20), 1399-1400.

McKinsey & Co. (1995). *Connecting K-12 schools to the information superhighway.* Palo Alto, CA: Author.

Mills, E. (1996, August 18). Vendors exploit electronic commerce tax puzzle. *InfoWorld.* Dow Jones News/Retrieval.

Musgrave, R. A., & Musgrave, P. B. (1989). *Public finance in theory and practice* (5th ed.). New York: McGraw-Hill.

National Bellas Hess, Inc. v. Dep't of Revenue, 386 U.S. 753 (1967).

National Defense Education Act of 1958, Pub. L. No. 85-865.

Picus, L. O., & Crampton, F. E. (1997, March). *Revisiting the sales tax: Opportunities for revenue enhancement at the state and federal levels.* Paper presented to the annual conference of the American Education Finance Association, Jacksonville, FL.

Quill Corp. v. North Dakota, 504 U.S. 298 (1992).

Rothman, D. H. (1996, August 21). Real books on the Internet. *Washington Post,* p. A25.

San Antonio Independent School District v. Rodriquez, 411 U.S. 1 (1973).

Sandberg, J. (1996, June 17). Making the sale: The allure of on-line commerce, its proponents argue, will eventually prove overwhelming. *Wall Street Journal.* Dow Jones News/Retrieval.

Sherman, S. (1994, April 18). Will the information highway be the death of retailing? *Fortune.* Dow Jones News/Retrieval.

U.S. Department of Commerce. (1995). *KickStart initiative: Connecting America's communities to the information superhighway.* Washington, DC: National Telecommunications and Information Administration, U.S. Advisory Council on the National Information Infrastructure.

U.S. Department of Education. (1995). *Projections of education statistics to 2005.* Washington, DC: National Center for Education Statistics. Office of Educational Research and Improvement.

U.S. Department of Education. (1996, June). *Getting America's students ready for the 21st century: Meeting the technology literacy challenge.* Washington, DC.

U.S. Department of Education. (1997, January). *Statistics in brief: Advanced telecommunications in U.S. public elementary and secondary schools, Fall 1996.* Washington, DC: National Center for Education Statistics.

U.S. Department of the Treasury. (1996, November). *Selected tax policy implications of global electronic commerce.* Washington, DC: Office of Tax Policy.

U.S. General Accounting Office. (1995, April). *School facilities: America's schools not designed or equipped for 21st century.* Washington, DC: Author.

APPENDIX 5.A

Calculation of Conservative Scenario Estimates

In the conservative scenario, lost sales tax revenues from existing electronic merchants will increase 10% annually, from 20% of the revenue base in 1997 to 50% in the year 2000, assuming a 5% sales tax. The portion of revenues lost due to the migration of traditional in-state merchants to multistate electronic merchants will increase .25% annually, from .25% of sales tax revenues in 1997 to 1% in the year 2000.

Please note that Table A.1 dollar amounts are expressed in millions whereas Table A.2 dollar amounts are expressed in billions.

TABLE A.1 Lost Sales Tax Revenues From Existing Electronic Merchants

Year	Revenue[a] (millions)	Sales Tax Rate[b]	Percent Interstate	Tax Revenue Lost (millions)
1997	950	.05	.20	9.5
1998	2,300	.05	.30	34.5
1999	4,500	.05	.40	90.0
2000	6,900	.05	.50	172.5

a. Revenue data based on projections by Forrester Research (1996).
b. Sales tax data based on median state sales tax rate for 1995, from U.S. Census Bureau.

TABLE A.2 Lost Sales Tax Revenues Due to the Migration of Traditional In-State Merchants to Multistate Electronic Merchants and Total Lost Sales Tax Revenue

Year	Sales Tax Revenue[a]	Percent Interstate	Tax Revenue Lost (Billions)	Total (from Table A.1) (billions)
1997	132.2	.0025	.330	.340
1998	132.2	.0050	.661	.695
1999	132.2	.0075	.991	1.081
2000	132.2	.0100	1.322	1.490

a. Sales tax data based on total general state sales tax revenues for 1995, from U.S. Census Bureau.

APPENDIX 5.B

Calculation of Worst-Case Scenario Estimates

In the worst case scenario, lost sales tax revenues from existing electronic merchants will increase 10% annually, from 50% of the revenue base in 1997 to 80% in the year 2000, assuming a 5% sales tax rate. The portion of revenues lost due to the migration of traditional in-state merchants to multistate electronic merchants will increase 1% annually, from 1% of sales tax revenues in 1997 to 4% in the year 2000.

Please note that Table A.3 dollar amounts are expressed in millions whereas Table A.4 dollar amounts are expressed in billions.

TABLE A.3 Lost Sales Tax Revenues From Existing Electronic Merchants

Year	Revenue[a] (millions)	Sales Tax Rate[b]	Percent Interstate	Tax Revenue Lost (millions)
1997	950	.05	.50	23.7
1998	2,300	.05	.60	69.0
1999	4,500	.05	.70	157.5
2000	6,900	.05	.80	276.0

a. Revenue data based on projections by Forrester Research (1996).

b. Sales tax data based on median state sales tax rate for 1995, from U.S. Census Bureau.

TABLE A.4 Lost Sales Tax Revenues Due to the Migration of Traditional In-State Merchants to Multistate Electronic Merchants and Total Lost Sales Tax Revenue

Year	Sales Tax Revenue[a] (billions)	Percent Interstate	Tax Revenue Lost (billions)	Total (from Table A.1) (billions)
1997	132.2	.01	1.322	1.346
1998	132.2	.02	2.644	2.713
1999	132.2	.03	3.966	4.123
2000	132.2	.04	5.288	5.564

a. Sales tax data based on total general state sales tax revenues for 1995, from U.S. Census Bureau.

SIX

Teachers and Parents Without School Buildings

HOW TEACHER UNIONS CAN USE TECHNOLOGY CREATIVELY

F. HOWARD NELSON

Education policy is set by state law or local school boards. Teachers have almost no say over standards for entry to the teaching profession or the hiring and evaluation of teachers. Most collective bargaining laws specifically restrict authority for educational and professional issues to management. Even though teachers in a typical public school have little say over what goes on in schools, parents nevertheless evaluate teachers by the schools they teach in. Parents are as likely to blame teachers as school boards or administrators for problems in their schools.

Advances in the creative use of technology, especially the World Wide Web, offer teachers a way to link directly to parents without the layers of state policy, school board control, and administrative interference that insulate teaching professionals from parents in schools. Similarly, teacher unions can break free from the narrow role assigned to them by collective bargaining laws and contribute directly to help-ing families prepare their children for school. The Web also enhances the ability of teachers and parents to organize on issues of common interest.

In the school building itself, teachers are not readily accessible to parents, even for simple phone calls, in part because most teachers do not have telephones. The working hours of parents align closely with school hours, making school visits difficult. Principals zealously re-strict parents' access to teachers and they are more concerned over what teachers tell parents than what teachers tell students. School buildings are also the fundamental unit of the teacher union structure. Faculty mailboxes and bulletin boards are critical to union organizing. The "building representative" is the first-level elected union leader-ship position.

Reflecting the decentralized aspects of U.S. schooling, organized teachers do not have a single Web address. Unlike Microsoft, Pizza Hut, or the White House, which have a single international Web address, the Web presence of teacher unions reflects the federated structure of teacher unionism. The two major national teacher unions contain about 90 state education associations or federations. And then there are the 10,000 local teacher unions, and dozens of them already have Web addresses. The potential number of teacher union Web sites guarantees no single vision of how the Web should be used and what should go on it. This chapter outlines three ideas relating to parents: (a) parent areas on union Web sites providing direct assistance, (b) union image building, and (c) involvement of parents in community organizing.

Parent Areas on Teacher Union Web Sites

With the help of volunteers, many local unions already provide homework help for students and parents through chat rooms or other on-line connections. Unions only do a fraction of the homework help business. A Web search on "homework help" yields dozens of sites, many with some kind of commercial angle. Many sites use volunteer teachers. Union homework hotlines aimed at parents are likely to be most useful if targeted to specific local issues such as preparation for district-specific examinations or state-mandated tests or if tied to specific components of the school district curriculum. Another fruitful model is business sponsorship of homework hotlines in which corporations pay union-recruited teachers to work after hours.

A part of local, state, and national union Web sites could be designated as "parent" areas. The focus should be much broader than just parent involvement. Content for parent areas already exists in union-run professional development programs; the archives of union publications; and numerous surveys, reports, and position papers. The hundreds of potential topics include appropriate early childhood education, what readiness for school means, how to communicate with adolescents, what your child should know at each grade level, how to read to your child, how to teach math facts, and how to integrate phonics and whole language.

Parent-oriented material on education and schooling already exists on the Web from governments, universities, commercial vendors, foundations, individuals, and even the White House, but teachers could add their unique imprimatur. The difference is like that between getting information on toothpaste from the American Dental Association or from Colgate-Palmolive. Endorsement of products and services, a sort of clearinghouse or *Consumer Reports* activity, is also possible. Former American Federation of Teachers (AFT) president Al Shanker often used his *New York Times* column, "Where We Stand," to promote or debunk specific education policies or ideas.

Teachers could help parents with "process" help, including on-line training, aimed at improving parent-teacher communications and parent-school relationships from the teachers' perspective. Parent involvement sites on the Web already cover some of this material, including Project Appleseed and the Alliance for Parental Involvement in Education. Topics could range from what teachers expect from parents and what parents should demand of teachers to information on religious rights in public schools and tips for working with principals, school boards, and teacher unions to get what parents need. Similarly, on-line training activities could address such topics as how to be good school volunteers, chaperoning on field trips, and raising money for your school. Local teacher union Web sites could get very specific with this kind of information. Newsgroups, listservs, and chat rooms could be used to add interactively to parent areas of union Web sites. These areas need regular monitoring by teachers, however. There is little purpose in having parent sites on union Web pages if parents only talk to each other.

The Channel One for Teacher Unions

Conservatives have complained about the liberal press for years, but when it comes to teacher unions, the *Washington Post* looks like the *Wall Street Journal* and the *Journal*'s reporting on teacher unions belongs on its editorial page. If this perspective sounds paranoid, then it fits well with that of other "disenfranchised" groups that have turned to the Web. One way to counteract the journalistic blackout is to offer parents a direct pipeline to the union, including information on union leaders and staff, the purpose of unions, and union positions on education and noneducation issues (ranging from family leave to child labor in Pakistan).

The knee-jerk reaction of teachers to commercial advertising on the Web, especially the subtle and sophisticated home pages for specific products and corporations, is to restrict children's access to these sites (see Pasnik, 1997). Many corporate sites are designed to bypass adult authority. Games and prizes encourage participation. Some companies use on-line sites to collect data on potential customers. Instead of just complaining, teachers should learn. When President Clinton stopped the pilots' strike against American Airlines early in 1997, the Web page of the pilots' association spoke to member interests on contract issues. American Airlines used its Web page to sell tickets and address issues important to consumers, not stockholders.

It would be a mistake, however, to use the Web merely as a billboard for union positions or a forum for complaining about mistreatment. Instead, the soft sell on unions should focus on the interests of children and on featuring union members to the public. Union print publications aimed at their own membership already do a good job of this. They just need to go public. Given the stereotypes about unions as organizations run by old white men, the qualifications and diversity (age, ethnicity, and gender) of union political leaders and staff should be showcased. Labor history and teacher union history, including photo essays, could be included. Factual data should be presented on such matters as union dues, budgets, and political contributions. The best information on labor laws, collective bargaining, strikes, tenure, and like subjects should be on union Web sites.

Organizing Support for Issues Important to Teachers and Parents

The Web offers opportunities far beyond passive statements on issues important to teachers. The restrictive nature of collective bargaining laws keeps important educational issues off the bargaining table. Lobbying in state capitals is one way to influence educational policy. Much of education is also a local matter, and community organizing has become an important part of union work. Parents are the most important allies in community organizing, and the Web offers new opportunities for issue education, materials distribution, training, and communication.

The AFT launched the Lessons for Life: Responsibility, Respect, Results campaign to refocus the public's attention on solutions to the problems facing public education and away from the agenda of abandoning public schools. Safe schools, better discipline, and higher educational standards undergird the campaign. Lessons for Life is a good example of a national campaign to develop and implement education policy primarily at the local level. The Web site of the national campaign contains information on campaign issues, prominent individuals endorsing the campaign, and background materials. Aimed as much at teachers as parents and community groups, however, the Lessons for Life Web site falls short of its community organizing potential in the areas of materials distribution, training, and communication.

Materials. Community organizing requires position statements, background research, fact sheets, speaking points, press materials (press releases and op-eds), and newsletters. Printed materials are expensive to make and distribute. Increasingly, information is needed in electronic format for easy reformatting for local needs and slants. At almost no cost, detailed materials can be posted on the Web and updated at any time. Users then download what they need rather than getting huge packets of information in the mail. Electronic downloading allows easy rewriting and graphic design. Instant access to Internet documents beats overnight mail. Directions and training should be an integral part of materials distribution.

Training. Community organizing requires training on specific issues and techniques of community organizing. Training on strategic matters is probably not appropriate for the Web, but the Web can be used for issue education and generic community organizing (choosing a good issue, how to deal with school boards, how to run a meeting, etc.). Some grassroots organizing training is already on the Web at such sites as the Electronic Community Organizing Web Page and the Citizen's Handbook, which cover topics such as planning and acting, getting noticed, getting and keeping people, facilitating, and fund-raising. On-line training manuals are more like materials than training, but more sophisticated tutorials and interactive training are possible (Filipczak, 1996). The Web can also supplement traditional workshop training by providing syllabi and handouts to go with live training, much as the papers presented at some academic professional meetings are posted on a Web site.

Communication. One problem with the AFT's Lessons for Life campaign has been getting feedback from individual districts and schools. A need also exists to share success and failure stories. These needs and problems are endemic to community organizing everywhere. One way to get feedback and cross-fertilization is through newsgroups, listservs, and chat rooms.

Grassroots Political Support. Opponents of public education have been very effective in mounting letter writing, fax, and phone bank campaigns. Teacher unions too have started to capitalize on such technology as fax hotlines, listservs and Web-based "fax sites," but the use of this technology has been confined to union members, leaders, and staff. Florida Education Association United, for example, has a "Power Fax" area on its Web site that enables teachers to identify and send faxes to their elected state and national officials. The Web offers an opportunity for unions to enable parents and other supporters of public education to rapidly find the telephone and fax numbers or e-mail addresses of state and national political leaders—even their local city council or school board members. Issue education and voting records should be part of this long-distance, grassroots effort.

Hope or Hype?

All of the ideas presented in this chapter can be implemented with existing technology and Internet architecture. Yet, most of these ideas probably seem a bit far-fetched, even counterproductive. Do teacher unions even represent the teaching profession and would parents bother to use union Web sites? Since most information on the Web is found through subject matter searches, good content is all that counts. Most parents look for good information and not whether it is provided by the International Reading Association or New York State United Teachers. A more important concern is that only some parents have good access to the Web and an understanding of how to use it. Parents with access tend to be more affluent and educated. This demographic selectivity presents enormous opportunities for organizing support for public education, but the parents who need the most help in parenting, working with teachers, and gaining access to school officials are the least likely to get help via the Web.

Technology aside, a related problem is the continuing debate within unions about the extent to which teacher unions should deviate from a narrow focus on the economic interests of members and take up education matters and professional issues. The membership orientation of unions naturally leads to an initial reluctance to put much of anything on the Web for public consumption. Both the National Education Association (NEA) and the AFT developed restricted-access areas on America Online (AOL) before they developed Web sites, and discount-priced AOL access charges are offered as a member benefit. As in most organizations, high technology use correlates with youth, and union leaders are usually not young. It is difficult to get a college degree without exposure to e-mail and the Internet. A survey of new teachers in Dade County, Florida revealed that 90% had access to e-mail and 70% used the Web. As always, the issue of resources presents itself, including both the expense and rapid depreciation of hardware and the slowness with which staff reorient their work from traditional activities to new technologies.

Conclusion

This chapter focused on the creative use of technology, especially the World Wide Web, to offer teachers a way to link directly to parents without the layers of state policy, school board control, and administrative interference that insulate teaching professionals from parents in the school setting. This chapter outlined three ideas for linking teacher unions to parents: (a) parent areas on union Web sites providing direct assistance, (b) union image building, and (c) involvement of parents in community organizing. Other important ways teacher unions utilize new technologies, such as member communications, recruiting, marketing member benefits, and membership accounting, have not been addressed. No single teacher union Web site exists because the Internet architecture allows any of 10,000 local unions to set up its own site. Decentralization aside, the ideas presented in this chapter do not identify a blueprint for action because the opportunities presented by technology change so fast. The best plan is to head off in the right direction and continually improvise as the technology changes.

References

Print Media

Filipczak, B. (1996). Training on Intranets: The hope and the hype. *Training*, pp. 21-27.
Pasnik, S. (1997). Caught in the Web of online advertising. *Principal*, pp. 24-25.

Web Sites

Alliance for Parental Involvement in Education: http://www.croton.com/alapie/
Allied Pilots Association: http://www.alliedpilots.org
American Airlines: http://www.amrcorp.com
American Federation of Teachers: http://www.aft.org

Citizen's Handbook: http://www.vcn.bc.ca/citizens-handbook/
Colgate-Palmolive Corporation: http://www.colgate.com
Electronic Community Organizing Web Page: http://www.mtn.org/eden-elect/ archive/msg00025.html
Florida Education Association United: http://www.feaunited.org
National Education Association: http://www.nea.org
Pizza Hut: http://pizzahut.ca/northbay/m_class.html
Project Appleseed: http://www.members.aol.com/pledgnow/appleseed/index.html
White House: http://www.whitehouse.gov/WH/New/ECDC/

Curriculum, Training, and Local Development

SEVEN

At a Distance
USING TECHNOLOGY IN DISTRIBUTED LEARNING IN HIGHER EDUCATION

BARBARA Y. LACOST

ALAN T. SEAGREN

SHELDON L. STICK

Changes in the delivery of instruction in postsecondary institutions are being fostered by spiraling costs, decline in revenue, competitive markets, increasing demand, and diverse student demographics (Baker & Gloster, 1994). The shift is toward an instructional model in which students have access to a variety of resources made available by the faculty, whose role becomes one of a collaborator or a mentor in the learning process (MacKnight, 1995, p. 29). Traditional class discussions, in which time and space hold captive both the professor and the students, are becoming a vestige of the past in part because of asynchronous communication systems.

Telecommunications facilitate the transfer of knowledge and provide greater access to experts and specialists independent of their locations. Recent advances in transmitting and managing video signals and images, as well as the interactive capability of innovative media, offer an unlimited potential for the practical transfer of learning experiences. Computer networks make it possible to deliver instruction to every desktop in every office, computer lab, classroom, residence hall, and home. With terminals in different locations, physical space is replaced with an access path (MacKnight, 1995, p. 31). Furthermore, researchers (e.g., Jones & Smith, 1992; Kulik & Kulik, 1991; Sawyer, 1992) report that students supported by technology-mediated instruction require about one-third less instructional time than students using traditional lecture or textbook methods. In fact, Baker and Gloster (1994) report that productivity gains can occur through greater retention of material, more efficient use of students' time, easy access to group study, more comprehensive feedback to faculty, and organized self-assessment and self-pacing.

Extending education through multiple media is not without challenges. Program development considerations must include compatible telecommunications infrastructures; software at each site that enables the delivery and receipt of materials; mechanisms for registration, advising, record keeping, assessment, and grading; and ample access to computer terminals for students to complete the work. If the delivery area includes international distribution and students in several time zones, provision for course delivery and completion of tasks and projects unhampered by time constraints is essential.

The Department of Educational Administration, Teachers College, University of Nebraska—Lincoln prepares leaders for positions in both K-12 schools and in higher education institutions. The faculty are involved in several programs based on telecommunication advances.

Provided in this chapter are descriptions of three approaches used in the Department of Educational Administration that distribute education among students at varying distances from one another—distributed education in higher education and leadership, distance education for K-12 administration, and the Teachers College master's program. The approaches differ in philosophy, in delivery, and in complexity. Each approach is described, and considerations for further use are provided.

Distributed Education in
Higher Education and Leadership

In 1994, the Center for the Study of Higher and Postsecondary Education (CSHPE) at the University of Nebraska—Lincoln initiated a technology-based graduate education program intended to eliminate the barriers of time, space, and stereotyping through Internet-linked personal computers and computer groupware. The 3-year pilot program, with approximately 15 faculty at Guam Community College serving as the participants, has enabled the distributed education program designers to define what is needed to realize the full potential of this exciting educational strategy. Over the past 3 years, enrollees have come from a variety of backgrounds and from the offices across campus to both coasts of the continental United States. The program is now a fully accredited, 13-course graduate program that leads to a doctoral degree in educational leadership and higher education. Dubbed "distributed education" by the multidisciplinary faculty team, the vision of a networked approach to teaching and learning incorporated interactive learning teams of students and facilitators. These teams currently dialogue from various locations in the world and work together to achieve desired educational outcomes. This challenging approach uses a multipath distributed education model, which is a flexible methodology that makes both multiple learning resources and strategies readily available. Use of the model allows for cost-effective and barrier-free educational opportunities. Students around the world can participate; instructors can pool their talents across great distances; classrooms never close; and absences and tardiness are no longer issues for faculty or students.

Philosophy

Distributed education creates a new learning culture that facilitates high levels of formal and informal interaction among students and faculty. The culture is shaped by students' dialogue as they question, support, and challenge each other to seek solutions to meaningful problems. The philosophy of the overall program is that teaching and learning are based on a "collaborative learning-reflection on practice" approach. A guiding principle is that learning is the responsibility of the student with faculty facilitating the process through encouragement, guidance, and leadership. A major premise of the program is that students and instructors do not necessarily need to meet at the same time and place for high-quality instruction to occur. Knowledge is viewed as an interdependent construction arising from conversation; reflective thought becomes an internalized social conversation. Teaching is intended to facilitate, engage, and empower learners to fully participate in the learning conversations. Students respond to discussion questions related to course readings and engage in multiple concurrent interaction with peers. Responses are shared among all; the need for immediate response is eliminated, leading to a more focused, thoughtful, and productive interaction (Stick, Seagren, & Watwood, 1996).

The faculty believe that such learning is best conducted through interactive computer-based communications. Collaborative learning is considered essential to the CSHPE program and requires a commitment from each enrollee to participate in interactive computer sessions with other enrollees. Collaborative education assumes that knowledge is socially constructed and that learning is socially interdependent. This approach reflects Slatin's contention that the technological environment allows for a shift from the standard pattern of initiation/response/evaluation that dominates the traditional classroom (as reported in MacKnight, 1995, p. 35). Such an environment provides for alternative sets of relations among the students, between instructors and students, and among all group members and the material being considered. Traditionally, the relationship between thought and conversation is that people can talk with each other because they think. The contrary view is that people can think because they talk with one another.

Delivery Mode

In this program, distributed collaborative learning takes place through "virtual groups" of which members are widely distributed geographically. Distributed education brings students together from diverse backgrounds and environments but does not require that they leave their environments. The distributed methodology enables courses to be taught through virtual interaction and collaboration. The computer groupware provides for open interaction among group members. The groupware stores information in multiple databases that can be created by the instructors or adapted from the templates provided by the software developer. A typical graduate course may have 15 or 16 databases that fall into three general categories:

1. Databases describing course content and operating information and instructions
2. Databases that address teaching and learning issues, including a course library of selected readings and source documents
3. Databases that focus on participant information, including a faculty division, a student division, and an informal area in which students can share personal, nonrelated information

Two important databases used by the faculty and students in the Department of Educational Administration are titled Virtual Class Program and Virtual Class Meeting. The former includes program areas in which assignments and course materials and information are posted; the latter is the heart of the asynchronous communication program and provides the forum for discussion, presentation, and interaction among the students and between the students and the faculty.

Instructional Strategies

The programming of course content consists of questions designed to promote discussion as each student response is shared with other group members. The instructor's role becomes that of a facilitator. A typical one-semester course has four to five modules, each activated for presentations and discussions for approximately 2 to 3 weeks. A series of assignments for each module is presented. A group size of about 15 appears to provide the best balance of interaction magnitude and scope (Lindsay Barker, electronic communication, 1995). A key concept is "replication," the process that allows students to access the server to send and receive learning materials, messages, and assignments. Accessing the server through replication provides immediate access to all messages, comments, and assignments posted by others and, in turn, takes in all information from the student. This information is then accessed by other students during replication. Distance has little meaning in this context because the learning activities might take place in an adjacent room, in local homes and workplaces, or at many widely dispersed sites throughout the country or across the world. Teaching, program facilitation, and learning resources are provided by the Department of Educational Administration. Teaching and advising employ a range of printed materials, and electronic communications technologies are designed to guide students' studies in their home locations.

Team teaching is considered essential for faculty flexibility and optimum use of teaching resources. Each course offering has a team leader and a minimum of one other faculty member. The main task for the teams is to encourage and lead learning conversations among the students. Faculty pose questions, facilitate communication, and assess outcomes. Printed material, computer software, and audio-visual materials are utilized to create the maximum possible degree of interaction among students and between students and teaching team members. The primary telecommunications vehicle is on-line education designed to permit distributed collaboration.

Content of Distributed Education

A 3-year schedule drives the center's program in higher education and leadership. Students admitted to the program have completed master's-level material. The 3-year program format enables students to progress at reasonable rates without experiencing unsustainable study loads. A fixed schedule with a planned course sequence is required to accommodate the demands on faculty time and on limited technical facilities without a major investment or redistribution of resources. All courses are theme and project based with problem analysis and solution development at the core of the process. Program themes for Educational Leadership and Higher Education include academic leadership, business and finance, student affairs and student development, policy development and analysis, and college teaching and learning. A campus attendance is required, typically for two summer sessions of 10 weeks each, but other arrangements that allow for equivalent on-campus experiences may be negotiated. Adjustments have been made in that schedule (e.g., four 5-week summer sessions; summer session preceded by a spring semester or followed by a fall semester). This aspect provides students the opportunity to fulfill four goals associated with the program:

1. To engage in face-to-face interaction
2. To access resources of the total university
3. To complete research courses
4. To devote some time to full-time study

Materials Required

Computer groupware is the vehicle by which faculty and students maintain their collaborative approach to instruction and learning. The university provides the software to the student after the enrollment process is complete. The students must own or have access to a workstation with an Internet connection. A 486 IBM-compatible personal computer with a minimum of 16 MB RAM and at least 50 MB of available disk space is adequate; the student may opt for an equivalent Macintosh system. All students are required to have access to library facilities that have a collection appropriate for graduate studies in leadership and higher education. Access to the University of Nebraska—Lincoln libraries through the Internet and the local and regional computer networks is offered to the students.

Costs

Start-up costs for the initial program were less than $20,000. Multiple licensing by the Department of Educational Administration and the university keeps costs for software to a minimum. When students complete the program, the software is returned to the university and redistributed to other students. The additional costs associated with student attendance, personal displacement, and career disruption are less than with conventional on-campus programs.

Distance Education for K-12 Administration

A second program launched by the Department of Educational Administration is currently offered to residents of Nebraska and adjoining states who are seeking a doctor of education that is focused on K-12 administration. The distribution of education in this case relies heavily on two-way teleconferencing, one-way television, and e-mail interactions between and among instructors and students. Students in campus-based classes at the university and students across the state participate in the same courses. Through the cooperation of the Department of Continuing Studies at the University of Nebraska—Lincoln, sites and on-site technicians are secured; that department also governs the registration of students, the collection of fees, and the distribution of texts and materials. Students distant to the university drive to designated linkup sites for access to two-way teleconferencing or one-way television. In the latter case, a course in session on the campus is fed to the distant sites in which the participants can view the instructor and the instructor's visual aids. The instructor and on-site students are able to interact only through audio with the students at the distant site.

The use of e-mail as a telecommunication strategy is a recent addition to the program. Although an outgrowth of the distributed education program, e-mail is more informal and does not utilize a multiple database system for organizing material as does the groupware previously described. Used for a series of required doctoral seminars, this telecommunication strategy, of course, has no visual interaction and is a part of the offerings to the students of K-12 administration. Students collaborate, share work with one another, respond to specific predetermined issues, and submit materials directly to the professors teaching the courses as a part of their regular workload.

Costs

Students who access their learning experiences through either two-way teleconferencing or one-way television register through the Department of Continuing Studies and are assessed an additional fee for the programs offered but receive a reduction in campus fees. When these same students access learning through the electronic mail courses, registration is through the general university system with no additional costs other than costs associated with maintaining electronic mail accounts. Students are encouraged to seek a graduate account with the university and are urged to purchase a service that provides them with full access to the university library system. Overall direct costs are reduced by travel and time commitment.

Distributed Education in the
Integrated Master's Program

The Teachers College master's program, a collegewide effort launched in 1996, is an offshoot of the CSHPE program initiated in 1994 and described in the first part of this chapter. The content of course offerings is designed for the baccalaureate-degree teacher with no or little experience in graduate-level education and is steeped in practice-driven inquiry. The program holds promise of being a cost-efficient system for integrating master's enrollees into the college's six departments. The format of the program allows specific departments to retain control and sponsorship of master's programs while opening student access to key courses and a wider variety of professors. Faculty are drawn from all departments in the college. Students, who move through the early stages of the program as a cohort, are urged to apply for admission to a specific department after experiencing the program but before the completion of the 18-hour core of coursework. Through distributed education, the program facilitates an integrated exploration of three core themes: inquiry, educational organization, and curriculum. In the traditional programs, these themes were isolated by the narrow boundaries of specific academic disciplines and the related specialties of professors.

Delivery

Again, shared groupware is the delivery basis of the program; delivery piggy-backs on the CSHPE's telecommunications infrastructure and support staff. Interdisciplinary teams of teaching faculty, drawn from the faculty of the departments in the college, create the common core curriculum that is offered to all participants, regardless of area of interest. Faculty volunteer to teach the courses as an overload to their regular assignments.

Philosophy

The underlying philosophy of this program is twofold. First, young scholars, those who have received their bachelor's degrees in the past 5 years, are more likely to persist in their education if continued and sequential knowledge and skill acquisition are available to them. Second, the demographics of Nebraska often require teachers to be at school sites that are long distances from postsecondary education opportunities, and procuring continuous, sequential post-baccalaureate experiences is difficult. Although the program was created with the distant learner in remote areas of Nebraska in mind as a prime consumer, the developers have been surprised at the interest shown by students who are nearer to the university but have busy schedules and urban commutes. Furthermore, although the program is specifically marketed to the student who has recently (in the past 5 years of so) completed a bachelor's degree in teacher education, teachers who have been out of contact with the postsecondary environment for much longer periods of time are initiating contact and expressing interest in participating in the program. In essence, the offering of educational opportunity through this more viable methodology appeals to all teachers interested in advanced skill development and encourages them to consider options that they previously may have not perceived as accessible. Students enroll as general master's candidates and after 18 hours of instruction through the use of the groupware apply for admission to programs offered in the departments of educational administration, curriculum and instruction, or vocational education. The original intent in producing the program was to ensure participation by students in the remote portions of Nebraska in a continuous yearlong master's program at the university.

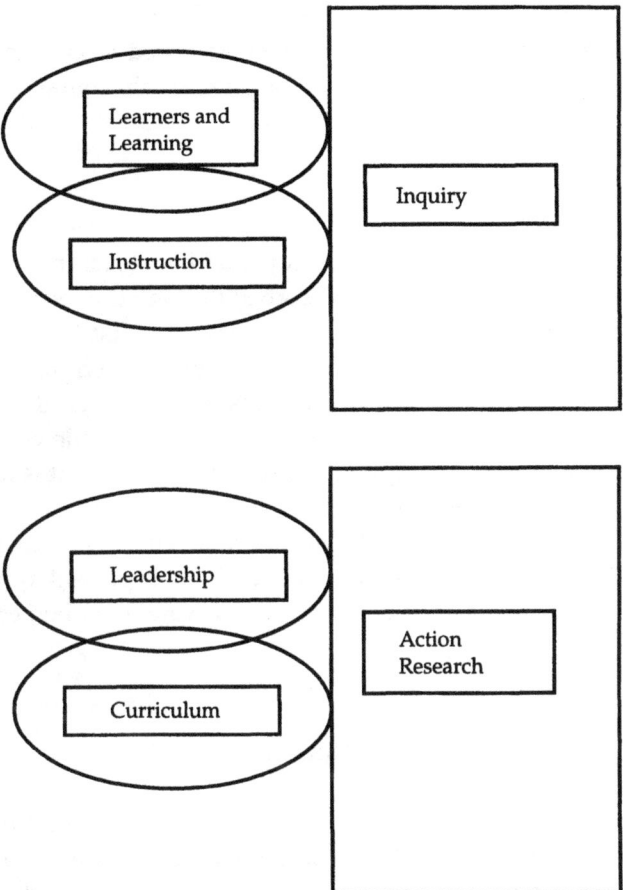

Figure 7.1. Content and Structure of the Four-Semester Teachers College Integrated Master's Program at the University of Nebraska–Lincoln

Content

The program consists of six courses, two of which focus on research. Each of the research courses requires a two-semester interaction and forms the foundation for the remaining four courses. Figure 7.1 displays the conceptual configuration for the coursework. Students progress from the study of learners and learning in educational settings through specifics related to instruction and curriculum issues. The skills and knowledge required of the experienced teacher in leadership settings are honed in a culminating leadership course.

Costs

All students pay the usual tuition and purchase texts and materials. The cost of software licenses are absorbed through the center, and the same procedure used by the CSHPE program for distribution to and subsequent collection of software from students is employed. User fees for campus access are waived; access fees through the Department of Continuing Education are added. The variable cost to the student is the personal computer. Again, access is all that is required for participation in the program. In some cases, computers that are purchased by employers of the enrollees are used; in other cases, students purchase new systems or upgrade their present systems to the requirements needed to support the software described earlier in this chapter.

Evaluation to Date

Evaluations indicate that the distance models are cost-effective as evidenced by the avoidance of costs associated with personal displacement, career disruption, and attendance. The programs, however, as with all newly implemented ventures, are not without challenges.

The same challenges that are present in traditional face-to-face learning encounters also are evident in this program. Keeping all students on the same topic, determining and maintaining closure of topics, increasing the pickup speed on new topics, and ensuring that students meet deadlines are issues with which faculty grapple on an ongoing basis in both the distributed and distance programs as well in the traditional classroom setting. Additional issues appear to be more incisive. For example, faculty workload is an ongoing concern. Development of university policy that attests that on-line courses constitute a class with the same status as on-campus face-to-face encounters is lagging behind the implementation schedule. Professors in educational administration, as well as in several other departments, generally teach the on-line courses as an overload; only occasionally does the telecommunication course substitute as a class-load course requirement offered in the traditional manner. In fact, in a few cases, professors are teaching on-line classes without compensation.

Advising students who are on-line and holding the required committee meetings also presents a challenge. The student's advisor spends perhaps an extraordinary amount of time creating a setting in which committee members may provide input and in which votes on specific issues can be made. Nevertheless, once a template is in place for the procedure, advising duties and issues are resolved. The groupware used in two of the three departmental distance offerings described seems a more viable solution; e-mail advising is somewhat unwieldy.

Residency requirements are expected to remain a continuing expectation of students. Nevertheless, adjustments must be made to accommodate those at great distances. The adjustments have included varying the required time spent in residence during summer sessions, allowing some type of field project that does not require the student to be in physical attendance at the university to serve as an extension of the residency, and implementing a procedure that provides for the completion of a required number of hours within a limited number of months to serve as a residency.

The opportunities offered through alternative distribution of education, on the other hand, transcend the challenges. An experienced team of university teachers has formed the opinion that the standards of student dialogue and conceptual understanding appear to be superior to those demonstrated in other teaching–learning experiences. For example, groupware learning may typically have in excess of 100 student-instructor or student–student interactions during a typical 3-hour graduate course, whereas the average for a traditional class might involve only 20 to 30 similar interactions (Lotus Development Corporation, 1997, p. 2). In fact, some of the on-campus courses are being redesigned based on the experiences with the groupware distributed model. Shared groupware expands the reach of the department and enhances its ability to attract a greater number of excellent students and experts. Use of the model has "allowed [the department] ... to accept some very bright students who may not have considered [the] program in the past" (A. T. Seagren, quoted in Lotus Development Corporation, 1997, p. 3).

The pilot experience of the faculty members in Guam demonstrated that faculty can best learn to use distributed education and understand its power when they experience it first as learners. Several students in the pilot provided their perspectives on involvement with the program. One student reported, "The process of writing down your responses forces a greater clarity of thought. Thus the responses of others are often more focused, creating an interaction that is more intellectually demanding of every student." Another suggested that

student interactions "provided me with information about authors that I had not heard of before. . . . Thus a broader range of ideas were brought to the class. . . . In many ways I have learned more from my colleagues by dialoging than from the actual course material." A third student related, "One of the greatest benefits is that you have time to set down and get into a deep discussion with another professional. In a traditional class, everyone runs in, takes notes, and then runs home." Students come to understand that learning is the sum of experiences, perceptions, knowledge, understanding, and problem solving.

Students in the master's program reported a variety of impressions. Students were open in reporting their initial ambivalence with the program and their subsequent satisfaction. One student related, for example, that she liked "to look people in the eye when I talk to them." She compensated by referring to pictures of students in a special file as she read and communicated with them. She went on to say, "It didn't take me long to get over those feelings; by the end I felt like I knew each person." Another student's comments focused on the process of taking responsibility for one's own learning. "I like the fact that I can work at my own pace when I need to. I do somewhat miss the classroom. . . . I didn't like to read their essays or responses until I had read the material and written my own response so that my response to questions would be my own work." Yet another student "thought the interaction among the students and with the professor was very beneficial. It allowed us to take learning to the next level. . . . You may never see the instructor or your classmates but you learn more about each person . . . because you communicate with them."

What Are We Learning?

What we are learning is that effective implementation of asynchronous communication as a learning strategy in postsecondary environments is and will continue to be dependent on cultural and organizational change. Goldberg and Richards (1995) declare, "As technology is integrated into all facets of [an] organization, the technology becomes . . . an impetus for change. The dynamic structure of a learning organization requires a technological infrastructure" (p. 15). And that structure must be more than the "one-to-many" distance techniques that have served as the cornerstone of technological integration in the past. Students who demand and respond positively to incorporation of multiple interactive strategies also serve as crucial change agents for our postsecondary environments. Evidence exists that such change is occurring in multiple institutions around the globe at varying paces. Institutions that envision the use of multidimensional interactive models in teaching and learning, that design policy and environments to support the programs, and that devote resources to the ongoing support and expansion of such programs will be the institutions that attract both capable and talented students and faculty.

References

Baker, W. J., & Gloster, A. S. II (1994, Summer). Moving toward the virtual university: A vision of technology in higher education. *Cause/Effect, 17*, 4-11.

Goldberg, B., & Richards, J. (1995). Leveraging technology for reform: Changing schools and communities into learning organizations. *Educational Technology, 35*(5), 5-15.

Jones, L. L., & Smith, S. G. (1992, January/February). Can multimedia instruction meet our expectations? *EDUCOM Review*, pp. 39-43.

Kulik, C. C., & Kulik, J. A. (1991). Effectiveness of computer-based instruction: An updated analysis. *Computers in Human Behavior, 7*(1), 65-94.

Lotus Development Corporation. (1997, June). Collaborative learning takes hold at the University of Nebraska. *Take notes! Lotus on campus and around the world* (Part No. 30939). Cambridge, MA: Author.

MacKnight, C. B. (1995, Spring). Managing technological change in academe. *Cause/Effect, 18,* 29-38.

Sawyer, W. D. (1992, January). The virtual computer: A new paradigm for educational computing. *Educational Technology,* p. 21.

Stick, S. L., Seagren, A. T., & Watwood, W. B. (1996, November). *Distributed education at the University of Nebraska—Lincoln.* Paper presented at the 21st annual conference of the Association for the Study of Higher Education, Memphis, TN.

EIGHT

Technology and Change in School Administrator Preparation

HANK BROMLEY

STEPHEN L. JACOBSON

This chapter is divided into three parts. In the first, Hank Bromley highlights certain social aspects of technological practice and develops a set of guidelines for thinking about technology use in educational settings. In the second section, Stephen Jacobson reviews his experiences with technology over the past decade as a faculty member involved with the preparation of school administrators at the State University of New York at Buffalo (UB). In the final section, Jacobson's experiences with technology in administrator preparation at UB are analyzed critically in relation to the criteria for technological usage developed by Bromley.

Although a sample of one limits our ability to generalize observations from one institution to others, we nonetheless feel that this case analysis of technology usage provides important insights into common misconceptions that may prove dysfunctional to administrator preparation and practice.

1. What to Ask About Technology
(Hank Bromley)

Technology has a peculiar status in our culture. Rather than approach technological activity as an ordinary pursuit, entailing the usual human concerns and frailties, we often view it as some sort of "fact of nature" that we must simply accommodate ourselves to. Educational decision makers, like others, often treat technology as somehow apart from the human world, instead of a fundamentally *social* phenomenon. The sort of judgment that would be exercised as a matter of course when any other new element entered a social situation (say, a classroom or district office) is usually suspended when the addition is a piece of technology.

This difference in approach can be seen in the following item, which appeared in a newsletter for school principals sponsored by Apple Computer. A box labeled "Hiring an Outside Consultant" (Bailey & Lumley, 1994, p. 6) offered suggestions for ensuring that engaging a consultant produced the hoped-for results. The ten suggestions all sound utterly commonsensical. Interestingly, though, if we replace the word *consultant* with *technology*, we generate suggestions that run counter to usual school practices. So, for instance, the first suggestion becomes, "Define your needs, then look for a [technology]," and the last becomes, "Ask yourself if the [technology] delivered what you expected." Although both pieces of advice ought to be obvious and unnecessary, neither practice described is at all common. Rather than starting with a determination of what we want schooling to accomplish and then examining how technology might be used to achieve those goals (the first piece of advice), computing initiatives are unfortunately most often based on the attitude "This technology exists, we've got to have it"—that is, educational computing has largely been technology driven rather than curriculum driven. As a result, putting computers in schools has all too often meant getting more of the same, only automated: electronic workbooks, computerized tracking of student "progress," and so on. And assessments of whether newly purchased technologies are living up to expectations are equally rare, also reducing the likelihood of computers contributing to meaningful educational improvements.

In this case, there is an a priori presumption that the addition of new technologies brings automatic benefits. Whereas other innovations—the use of paid consultants, for instance—are routinely planned with attention to the subtle hazards of vested interests, mixed motivations, unforeseen effects, and the like, the special status of technology renders it largely immune from such considerations. The prospect of "more technology for schools" is self-evidently desirable, and efforts to raise questions—even merely saying, "Use the same judgment here you would in any other situation"—are perceived as antitechnology statements. Because of the presumption of technology's largess, skepticism appears simply irrational and is therefore dismissed without consideration.

The "ordinary language" for technological evaluation remains impoverished precisely because the questioning of technology is one of the few topics still considered taboo in polite American society. Outside of small coteries of radical environmentalists or postmodernists, most attempts to criticize the roles of technology are answered with ad hominem charges of "Luddite," thus foreclosing genuine debate. (Crawford, 1994)

In such a context, a serious examination of what social visions are built into—and in turn enacted by—a given technology is hardly likely. But if one were committed to conduct such an examination, what should be considered? What questions ought be asked?

Technology as Carrier of Social Relations

One set of important considerations is highlighted by the recent popularity of *integrated learning systems* (ILS). Increasingly resorted to by urban school districts under pressure to show measurable improvements in student performance and swayed by vendors' assurances of consistency and all-in-one convenience, ILS combines presentation of material, testing, and tracking of student progress in one automated package. But it is a package with a severely restricted understanding of education. ILS "labs," equipped to process students by the roomful,

> are prime examples of the non-neutrality of technology. They do not foster all or even several types of learning but rather one particular, and particularly narrow, conception whose origin is not with teachers who work with children but with the technologists, industrialists, and military designers who develop "man-machine systems." They do not encourage or even permit many types of classroom organization but only one. They instantiate and enforce only one model of organization, of pedagogy, of relationship between people and machines. (Hodas, 1993)

The question raised by this example is, What sort of baggage comes along with a given technology? What assumptions are built into the technology (and imposed on its users) regarding the nature of learning, ideal classroom organization, the goals of schooling, and so on? In the case of ILS, administrators who find the technology very attractive because of the efficiency it promises may be unaware of its curricular implications; they may not recognize the student-as-cog-in-the-machine model it imposes on the classroom. (Or they may be sufficiently removed from classroom life to believe that model can feasibly be applied to the classroom.)

One might respond to these concerns by contending that whatever potential hazards may attend instructional use of a given technology, administrative uses are a separate matter. And to some degree they are. But administrative usage of certain technologies can nonetheless have widespread effects by generating a dynamic influencing the entire mode of administration.

As an example of how computer use can enforce a formalized, abstracted mode of social interaction, consider this experience of mine: A few years ago, I received a blank form from the publisher of a directory of households in my city of residence, with a request to fill in information pertaining to my household and return the form. The first line was marked "husband," the second was marked "wife," and the rest were designated for "other occupants over 18 years of age." My household at the time happened to consist of five unrelated adults. We had no husbands and no wives. I called the publisher to complain about the apparent assumption that all households contain a married couple. The person who answered the phone was pleasant but not particularly helpful. She suggested I cross out the labels "husband" and "wife" and fill in our names in any order. I asked how the information would then be entered into the publisher's database. She said that was no problem, whoever happened to be on the first line would be labeled head of the household and we would be listed in the directory under that name. Which is to say, I could cross out whatever I liked, but it would have no effect on what got into the database. I told her our household had no head, and we would not be returning the form.

Had their listings not been standardized into a fixed format with everything stuffed into an abstract set of categories, had their final product simply been a large sheaf of cards users could flip through, then I could have scribbled any sort of explanation on our card and it would remain there for users to see. But in formalizing the information, in preparation for computerizing it, they conclusively imposed an ordering on it, one the exclusions of which would not necessarily be obvious to users of the directory, who would just see a tidy list of households, organized "naturally" by name of household head.

In a similar experience during a recent job search, I received "affirmative action" forms from most of the universities I applied to work at, requesting that I volunteer information on my sex, race, veteran status, and so on, so the university could gauge its success at reaching goals of diversifying its employee pool. One of the by-products of this particular method of pursuing those laudable goals is that applicants' identities are classified into arbitrary, fixed categories. For instance, one such form, under the "Racial/Ethnic Data" section, asked me to choose which of five groups I identify with (American Indian or Alaskan Native; White [not Hispanic]; Black [not Hispanic]; Asian or Pacific Islander; and Hispanic), noting, "We can record only one racial/ethnic choice; if more than one is chosen, it will be recorded as *unknown.*" Now, what if I had one black and one white parent, or one Asian and one black parent? Or parents who were themselves bi- or multiracial? Or if I affiliated culturally with more than one group for other reasons? In response to the imperative to count the members of various categories and abstract the entire applicant pool into a single set of numbers, this university (like many others) formalized racial identity in a manner that excluded many possibilities and rendered the exclusions invisible to users of their data, just as with the city directory example. But what's worse, in this case an initiative specifically intended to welcome a broader range of people into an institution effectively tells many of them they don't even exist.

Similar constrictions of thought can arise from the use of highly formalized information systems in school settings. Consider the popular Instructional Management System marketed by Abacus Educational Systems. This performance-monitoring and report-generating software package is intended to assist with curriculum design, lesson planning, test generation and scoring, classroom-level record keeping, and building and districtwide performance evaluation. It integrates all these tasks via basing instruction on lengthy lists of simple, specific objectives. Student progress is continually monitored on a check-off basis, "yes" or "no" on each objective, and these results are readily aggregated to any level of interest, at any time. But if adopted, the Abacus package does far more than "assist" with these pedagogical and administrative tasks; it will in fact determine important aspects of the educational process by constraining the form of instructional objectives. The content of the objectives may be freely specified by each district—so long as the objectives are uniform across the entire district and student mastery of each objective can be expressed as a simple "yes" or "no," as determined by computer-scored, multiple choice tests.

The adoption of such a system clearly creates enormous pressure for adhering to certain educational philosophies rather than others. What if, for instance, a teacher felt the most important thing for students to learn was how to ask good questions? Where would that fit into this scheme? And the system clearly also favors certain modes of leadership and kinds of relations between teachers and administrators.

The Abacus system is understandably attractive to many upper-level administrators, because it addresses, in a seemingly precise fashion, pressures they face to provide "accountability" for district performance. But in turn, the operation of the system passes those same pressures along to subordinate administrators and classroom teachers. In other words, the Abacus system is a piece of technology that is shaped by particular social pressures so as to embody particular views on how schooling ought be conducted and then constrains users of the system to act in accordance with those views.

One of the social forces shaping the Abacus system is a reverence for the value of information per se. In this regard, Abacus is representative of a much wider cultural phenomenon. In the introduction to his book *The Cult of Information*, Theodore Roszak reflects on the contemporary enthrallment with information:

> The word has received ambitious, global definitions that make it all good things to all people. Words that come to mean everything may finally mean nothing; yet their very emptiness may allow them to be filled with a mesmerizing glamour. The loose but exuberant talk we hear on all sides these days about "the information economy," "the information society," is coming to have exactly that function.... People who have no clear idea what they mean by information or why they should want so much of it are nonetheless prepared to believe that we live in an Information Age. (1986, pp. ix-x)

The ubiquitous talk of an impending information age suggests that benefits will accrue automatically; most of the information age rhetoric invokes an image of vast improvements in civic participation and access to resources, brought about by the mere presence of new technologies. Langdon Winner lists some of the key fallacies underlying these claims in his essay "Mythinformation": (a) people are bereft of information, (b) information is knowledge, (c) knowledge is power, and (d) increasing access to information enhances democracy and equalizes social power (1986, p. 108).

To suggest that participation in public life is currently limited by inadequate amounts of information is misleading. Although some specific kinds of potentially helpful information are not well distributed, on the whole people are drenched in information. The problem isn't getting enough, but making sense of what's available; providing everyone with an on-ramp to the information superhighway won't help with that problem. Information—raw data and facts—does not amount to knowledge until it is organized somehow, shaped by an intelligence, gathered toward some end. And knowledge does not constitute ideas—let alone wisdom—until it is further digested and pondered. Ideas may in some sense be power, but knowledge is not, much less information. Ideas are what help people make sense of public events. There are plenty of raw data about, so many in fact that they inhibit our ability to perceive and grapple with the operant ideas:

> When we blur the distinction between ideas and information and teach children that information processing is the basis of thought . . . we bury even deeper the substructures of ideas on which information stands, placing them further from critical reflection. For example, we begin to pay more attention to "economic indicators"—which are always convenient, simple-looking numbers—than to the assumptions about work, wealth, and well-being which underlie economic policy. Indeed, our orthodox economic science is awash in a flood of statistical figments that serve mainly to obfuscate basic questions of value, purpose, and justice. (Roszak, 1986, pp. 106-107)

In one sense, Abacus is simply another manifestation of this tendency. Just as fundamental questions of how and why wealth is distributed are displaced by preoccupation with statistical minutia, so are fundamental questions of what education is for displaced by preoccupation with an enormous stream of numbers. Although the computer has intensified the problem, it certainly did not create it. The power of low-level facts to sway public opinion derives from a worldview of which the computer is simply the latest incarnation. "Behind the style stands the mystique of scientific expertise that lends authority to those who marshal facts in a cool, objective manner. The computer is simply a mechanical embodiment of that mystique" (Roszak, 1986, p. 164).

Information is no substitute for the ideas that enable understanding of the social world. Nor does it suffice to enable affecting the world; that capacity depends more on organized action than on information. "The formula information = knowledge = power = democracy lacks any real substance. At each point the mistake comes in the conviction that computerization will inevitably move society toward the good life. And no one will have to raise a finger" (Winner, 1986, p. 113).

The Importance of Context

Alongside the question of what assumptions are built into a given technology, what values it embodies, another equally important question to consider is the context in which the technology is used. Much literature on educational technology has a narrow focus on the characteristics of the technology itself. A full understanding of why a particular piece of technology is or is not used, or why it is used in particular ways or has a particular impact, is unlikely to be achieved without careful attention to the social context of its use: Who is using it, why, toward what ends, under what conditions and pressures? All of this has as much to do with the eventual outcome as the nature of the technology itself does.

I am currently involved in a research project at a local school, where it has become apparent—to no one's surprise—that such contextual factors as an unfavorable student-to-teacher ratio help determine the mode of classroom computer use. The staff at this particular school are committed to fully integrating the computer into the curriculum rather than treating it as some sort of extraneous add-on. But given that there are too few adults in the classroom to meet the diverse needs of the students, teachers sometimes find they must resort to using the computer as a reward for good behavior and withdrawing access as a punishment for noncooperation. In the immediate situation, the teacher gets compliance and the student gets some experience with the technology. But ultimately, this practice interferes with integrating the computer into the curriculum, and students become habituated to a carrot-and-stick model of social interaction—even though no one involved seeks these outcomes.

A study by Michael Apple and Susan Jungck (Apple, 1993, chap. 6) draws similar conclusions. It shows how the day-to-day realities of teachers' lives lead a conscientious but overworked set of professionals to employ an utterly routinized and vapid computer curriculum, simply because it was already prepared (freeing them from having to write one) and kept the students busy (freeing the teacher to complete other tasks). Unfortunately, the least intellectually engaging instructional software (such as drill-and-practice programs) can become the most attractive to overstressed teachers, because it keeps students wholly occupied, in a known activity with few surprises to require the teacher's attention, for a predictable amount of time.

From Whose Perspective?

Another consideration often overlooked is how people in different social positions can have very different experiences with the same technology. Rather than ask whether a particular use of technology is a good idea, we need to ask, "Good for whom?" Who benefits (and in what ways), and who doesn't?

In the case of instructional use of computers, broad statistical portraits consistently disclosed systematic inequities throughout the 1980s and early 1990s (see Sutton, 1991, for an extensive research review). Although these inequities may have begun to diminish in recent years, that they persisted for so long—and in some respects still do persist—indicates definite blind spots in our thinking. Throughout this period, measurements of computer use both in and out of school for students at all ages, in several countries, found less access for girls as well as for students of color, children from low-income families, and students labeled "low ability." But more to the point in the present context, the type of use varied along the same lines. Even when students from these groups were provided access to computers, they were disproportionately engaged in drill-and-practice software, "mastery" learning of decontextualized basic skills, and vocational training in the use of specific software, whereas boys, white students, middle-class children, and students labeled "high ability" were disproportionately involved in open-ended simulations, integrated applications, and programming.

In effect, some students were learning how to direct the new technology while others were learning how to be directed by it. Sutton (1991) argued that this differential reflected the application of existing school practice and ideology to use of the new technology, the dominant views being "that children must first master the basics before moving to higher order thinking and that poor and minority children lack the basics" (p. 482). As a result, the already advantaged became more so, adding yet another domain to their list of advantages. The computer, introduced partly in hopes of creating new opportunities for all children, by and large made things worse, even when everyone got to use it.

The Abacus system discussed earlier provides another example of a single technology having notably different effects on different people, according to their structural location. In this case, a tool that can ease the burden on some administrators can simultaneously hinder the work of other administrators and of teachers.

Regardless of how the lines are drawn, there are always varying need and interests; it is consequently always necessary to "disaggregate" the question of a given technology's impact.

Key Questions

Based on these examples, we can propose a set of questions that ought to be asked if one is serious about understanding the meaning of any technological intervention in schools:

1. Why is this initiative even occurring? In particular, is it technology driven (based on a perceived need to have the latest technology) or curriculum driven (based on a careful discussion of educational goals and of what means are lacking to reach those goals)?
2. What social visions are built into—and in turn enacted by—a given technology? Does it enforce particular forms of pedagogy or of classroom and school organization? Does it impose a certain conception of knowledge or of the learning process? Is it compatible only with particular views of what schooling is for or consistent only with particular sets of relations among school personnel?
3. How is the context of use likely to shape the way this technology is employed? Who is using it, why, toward what ends, under what conditions and pressures, with what supporting resources?
4. Disaggregate the impact (i.e., do not limit your view to the effects on the most visible or most powerful persons): How are groups of people in different structural locations likely to be affected differently by this initiative? Who will be helped, and how? Who will be harmed, and how?

2. Reform and Technology Use in Administrator Preparation
(Stephen L. Jacobson)

In this section, we review recent calls for changes in administrator preparation and the evolution of the use of advanced electronic technologies, especially microcomputer technology, in the preparation of school administrators. We use, as the basis of this discussion, the experiences of one of the chapter's coauthors, Stephen Jacobson, a faculty member in the educational administration program at UB since 1986.

As we shall see, although recommendations for change in administrator preparation programs are fairly explicit with regard to such aspects of preparation as the quality and demographic composition of the candidates who should participate in such programs, the role of school districts in candidate selection, and the level of collaboration that should exist between universities and school districts, they are less clear about the role of technology. The resulting social vision implicit in educational administration's technological evolution suggests that, much like their counterparts in teacher preparation, university faculty charged with preparing school administrators have come to view technology as being more than just a positive influence on social change, rather they view technology as a primary moving force behind it (Carr & Bromley, 1997). The danger inherent in this type of technological determinism is that it ultimately ascribes change agency to technology and not to people (Carr & Bromley, 1997). If, during their preparation, future school leaders come to believe that change is not within their control, this misconception may relieve them of a sense of felt responsibility for promoting productive social change. Instead, these novice administrators may come to believe that they just have to keep their schools abreast of the latest technology, from which productive outcomes will follow.

The notion of placing technology at the epicenter of change contrasts sharply with our own beliefs. We agree with Fullan's (1993) assertion that meaningful change in education can be realized through collective action predicated on moral purpose and individual responsibility. First and foremost, educators must never lose sight of education's moral purpose: "The moral purpose is to make a difference in the lives of students regardless of background, and to help produce citizens who can live and work productively in increasingly dynamically complex societies" (Fullan, 1993, p. 4). Although we do not view technology as an essential prerequisite for change, we recognize fully that it is *not* a neutral tool: "Where the uses of new technologies are concerned, even though 'the future is unwritten' [Sterling, 1993], and to be determined by human action, some futures are definitely easier to bring about than others" (Carr & Bromley, 1997, p. 20).

What this means is that school leaders should consider the appropriate role of advanced technologies in bringing about desired changes, but only after contemplating whether these goals will make a difference in students' lives and help them to live and work more productively. "Students" in this regard can be read either as youngsters in elementary and secondary study or as adults aspiring to administrative positions through preparatory programs and practice. "School leaders," in the context of the next section of this chapter, should be read as the faculty responsible for the preparation of these aspiring administrators.

Reforming the Preparation of School Administrators

Close on the heels of the major educational reports of the 1980s (see, e.g., *A Nation at Risk* [National Commission on Excellence in Education (NCEE), 1983]; *Tomorrow's Teachers* [Holmes Group, 1986]; and *A Nation Prepared: Teachers for the 21st Century* [Carnegie Forum on Education and the Economy, 1986]) came *Leaders for America's Schools* (University Council for Educational Administration [UCEA], 1987), the report of the National Commission on Excellence in Educational Administration. The earlier reports criticized both the quality of candidates entering teaching and the preparation these individuals were receiving. As a result of reforms arising from these reports, school administrators found themselves labeled as poorly prepared, managerial rather than leader oriented, resistant to change, and confrontational as to teaching and teachers. Therefore, in *Leaders for America's Schools*, attention was shifted to the preparation of school administrators based on the assumption that better-qualified, better-prepared teachers will only be effective if the individuals charged with administering schools are also better qualified and better prepared:

The evolution of reforms over the past few years has progressed from cosmetic changes in course requirements to radical restructuring of the school environment. The new roles envisioned for teachers in reports of both the Holmes Group and the Carnegie Task Force on Teaching as a Profession draw education into a broader field of management research from which it has been isolated for too long. At the same time, these reports identify the unique setting of the school workplace, envisioning how teachers could respond to greater autonomy and professionalism. Yet, the reforms cannot be successful without strong, well-reasoned leadership from principals and superintendents. (UCEA, 1987, p. 5)

After examining preparation programs in the United States and Canada, the report identified "troubling aspects throughout the field," including

- Lack of a definition of good educational leadership
- Lack of leader recruitment programs in the schools
- Lack of collaboration between school districts and universities
- The discouraging lack of minorities and women in the field
- Lack of systematic professional development for school administrators
- Lack of quality candidates for preparation programs
- Lack of preparation programs relevant to the job demands of school administrators' work
- Lack of sequence, modern content, and clinical experiences in preparation programs (UCEA, 1987, pp. xvi-xvii)

Throughout the chapter, we comment on many of these legitimate concerns noted by the commission, but it is the last two deficiencies, that is, the lack of relevance to the demands of administrators' work and the lack of sequence, modern content, and clinical experiences that draw most of our attention, because these issues are of particular concern with regard to the use of technology within preparation programs such as UB's. We should note that although our focus in this chapter is with the use of technology in preparation, our primary concern is with the use of technology in practice. We feel that the preparatory experiences aspiring administrators have with advanced technologies should help to shape their understanding of the role of these technologies in school practice. In other words, by studying technology use in preparation, we hope to provide useful insights into the uses of technology in future practice.

Technology Use in Administrator Preparation

Since 1986, the educational administration program at UB has gone through a three-phase evolution. Phase 1 was simply talking about computers and their role in education; Phase 2 involved using computers as tools essential for solving basic administrative problems; and in Phase 3, the current phase, technology has become more or less transparent and is viewed as a medium for instructional delivery. This shift from technology as subject matter to technology as tool to technology as medium of instruction has been accompanied by a marked shift in the way courses are delivered.

During Phase 1, a traditional, instructor-led seminar approach was employed to study technology as subject matter. In the fall 1986 semester, when I arrived at UB, I was assigned Administrative Issues and Applications of Educational Small Computer Systems, a course first offered in spring 1984. As developed by my predecessor, the purposes of the course were threefold: (a) to comprehend the significance of microcomputer technology within an information society; (b) to comprehend the microcomputer's potentiality in educational administration for computer-aided office work and as a decision-support system; and (c) to analyze and synthesize policy issues and procedures for planning, implementing, and evaluating administrative microcomputer systems in education. Topics included conceptual bases for and administrative issues of computer system planning, implementation, and evaluation and administrative applications of computer systems.

My own use of computers at that time was limited to statistical analyses of data tapes on the mainframe and the use of databases, spreadsheets, and word processors on personal computers. Nevertheless, I was seen by my colleagues as a sophisticated computer user and therefore appropriately prepared to teach the course. When I inquired as to whether there were computers I could use for the course, I was informed that the course had, in the past, always been offered in a seminar conference room and that the computer labs were unavailable. When I pushed this issue further, I found that not only were computers not used in the course, they were not required for the completion of any course assignments, not even written assignments. Eventually, I was able to secure a little time for hands-on computer experience, but only on mainframe terminals. I made word processing of written assignments a requirement, only to find out that few students had regular access to computers, either at home or in school. Those who did have access to computers, usually the few administrators in the class, had, in most cases, given them to their secretaries because they considered computers a clerical tool.

This experience taught me that our students were not well acquainted with computers and thus had only a limited experiential base on which to consider the implications of technology use in schools. When the course was offered again the following year, I made it my business to secure a regularly scheduled time for the class in a microcomputer lab. I revised the course objectives: (a) to examine current issues surrounding the application of computer technologies in educational settings, (b) to anticipate the role of administrative information systems in school improvement, and (c) to introduce the student to administrative applications of microcomputers. Students were also required to give a 15- to 20-minute oral presentation in which they would demonstrate and evaluate a software package applicable to administrative use in an educational setting. My intention was to create a mixed approach in which the computer lab complemented the classroom activities, thus shifting the instructional emphasis from a purely didactic to a more hands-on approach. Unfortunately, the traditional structure of university coursework created a bit of a problem. Our courses are typically 3 hours long and offered once a week for 15 weeks. Although this format may work well for academic material, it is not as useful for sustaining the development of technical skills. Computer skills learned in the lab one week often had to be repeated the following week for those students who didn't have access to compatible hardware or software.

The next time the course was offered, it was moved to a hands-on training workshop during the summer session, 3 hours a day, 3 days a week for 4 weeks, all spent in a computer lab. This change marked the beginning of Phase 2, in which technology is used as a tool. In other words, from an instructional perspective, the focus of the course moves technology from the foreground to the background, with the emphasis instead being on basic problems of administration. To expedite the technical aspects of computer training, that is, learning to use the computer as a tool, Patricia Brosnan and I developed a tutorial program for the Macintosh (Jacobson & Brosnan, 1988). A computerized data analysis written on a word processor and developed through the use of additional software applications such as a spreadsheet, database, graphics, or statistical package was added to the course requirements. I also introduced a new text, Craig Richards's (1989) *Microcomputer Applications for Strategic Management in Education.*

In the first chapter, Richards (1989) notes that as a result of his own use of computers in policy analysis, he decided to change the orientation of his book "from a pedagogical focus on microcomputers and school business techniques to the strategic analysis of managerial problems more generally" (p. 1). This was perhaps the first text on administrative applications of computers in education to present the machine as a means to an end, rather than an end in and of itself. The earlier texts I had used for the course (Bluhm, 1987; Gustafson, 1985; Miller, 1988) focused primarily on the efficiency of computers for data processing and information management. Richards used instead a case study approach to highlight the role of the computer in decision support. The text assumes the reader has an understanding of basic statistics; some hands-on experience with spreadsheets, databases, and graphics packages; and an appreciation of the fact that two of the most significant responsibilities an administrator must face are policy analysis and decision making.

Highlighting the fact that educational policy is determined by strategic planning and management, which computers can facilitate, not merely by the selection of hardware and software, Richards (1989) provides analyses of critical issues of practice, such as grade inflation, declining enrollments, site-based budgeting, contract negotiations, teacher absenteeism, efficiencies in cash management and purchasing, program cost-effectiveness, and minimum teacher salary mandates. Contrary to the impression left by many of the earlier texts, technical issues do not supersede substantive matters. Using any one of the issues from Richards noted above, each student was required to develop a case study policy brief, and then, working in groups, the students had to develop and analyze a case study of their own making. There is a pervasive concern that computers tend to isolate learners and therefore are antithetical to collaborative learning. Yet, these group case studies suggested otherwise, because computers became a catalyst for student interaction. Group members would find themselves addressing fairly significant educational policy issues while gathered around a computer that was being used to input, sort, graph, and analyze relevant data.

Another example of technology serving as a catalyst occurred shortly after that summer session. Using abstracts of four chapters from *Reforming Education: The Emerging Systemic Approach* (Jacobson & Berne, 1993), an electronic salon was set up in advance of the American Education Finance Association's (AEFA) yearbook roundtable session held at the annual meeting in Nashville.[1] For 2 weeks prior to the Nashville conference, educators from around the world engaged in a lively discussion with the chapter authors and one another, using an electronic listserver established specifically for this purpose. After a slow start, the conversation heated up when a respondent from Senegal, after first describing the role of teacher unions in West Africa, raised serious questions about the potential of advanced technologies to further widen the gap between the developed and developing nations of the world. For approximately 4 days, responses came flooding in from all over the world, and during that time there was a sense that we had created a "virtual" community of international educators.

These experiences with technology as catalyst helped the transition from Phase 2, technology as tool, to Phase 3, technology as medium of instruction. Currently, in addition to computers, the faculty of educational administration is attempting to integrate advanced technologies such as electronic mail, the Internet, and interactive television into the delivery of instruction both on and off campus. For example, over the past year, two educational administration courses were taught using distance learning in cooperation with Project Connect, a consortium of approximately 10 interactive classrooms located in school districts and colleges coordinated by Erie 1 BOCES. Two more courses are planned for the upcoming year. In addition to distance learning, faculty members are experimenting with computerized presentations and simulations and electronic conversations and reading groups.

These Phase 3 attempts by faculty to infuse technology throughout the curriculum stand in marked contrast to the earlier phases, in which technology was treated much like other areas of academic study, such as school law and personnel administration, occurring primarily in a course designated specifically for that purpose. With time, this infusion of technology should make it relatively transparent; yet for the time being these same activities tend to bring the technology back to the forefront, often highlighting it at the expense of the subject matter.

3. Questioning Technology in Administrator Preparation (Hank Bromley and Stephen L. Jacobson)

In this final section, Jacobson's narrative in Part 2 is considered in light of the guidelines for thinking about technology use offered by Bromley in Part 1.

The chronology begins with Phase 1 and the initial version of the class Jacobson inherited in 1986 (see p. 141). This version of the class is vulnerable on all four criteria:

1. It is technology driven rather than educationally driven, rendering technology a solution in search of a problem. Its starting point is the presumption that the microcomputer ought to be used, and it then sets about demonstrating the computer's value in a series of task areas, rather than beginning with educational goals to be reached and asking what means can get us there.

2. It disregards the question of what social visions may be built into the technology and consequently color the conduct of the administrative tasks performed via the technology. The computer is presented as a neutral tool that may be employed toward any end, increasing efficiency without altering the outcome in any significant way.

3. It disregards as well the context of use and its influence on technological practice. The microcomputer is addressed in splendid isolation, idealized as functioning independently of the human agendas operative in any given setting.

4. It makes no effort to disaggregate the impact of computerizing the various administrative functions. It does not recognize how the retooled

procedures may, intentionally or otherwise, benefit some parties at the expense of others.

Jacobson's initial efforts to upgrade the class by including at least a modicum of hands-on computer practice proved frustrating, as most of the students lacked both experience with, and enthusiasm for, working directly with the equipment. The obstacles encountered might have been anticipated had more attention been given to Criterion 3, the entanglement of the technology with the ongoing social dynamics at its site of use. In this case, despite whatever suitability the microcomputer may *in principle* have had for use by current and future administrators in accomplishing various tasks, such factors as the difficulty of access and a pervasive notion that hands-on keyboard work was more appropriate for secretaries rendered it virtually impossible for the hoped-for outcome to be realized in this particular setting.

The administrator/secretary distinction also highlights the need to consider Question 4, on the differential impact of new technologies on various classes of people. Clearly, administrators and secretaries found their jobs altered in very different ways (at that historical moment) by the introduction of microcomputers to their shared workplace. Explicating the differences would, of course, be a considerable project in itself, but based on the outlook shared among Jacobson's students, one could reasonably conjecture that secretaries' workloads grew far more than did those of the administrators they assisted.

Phase 2, characterized by viewing the technology as a tool, involved shifting the format of the class to a hands-on training workshop focusing on a series of case studies and adopting the Richards (1989) text, which treated the computer as a means to an end (grappling with particular administrative issues) rather than an end in itself. In moving technology "from the foreground to the background," the new format and text ably address Criterion 1, concerning themselves with what administrators must accomplish and introducing technology specifically to assist with those tasks.

This version of the class does, though, remain vulnerable on Criteria 2 to 4. As described, the text and classroom exercises do not attend to the question of how reliance on microcomputers might shape the work of administrators, rendering some solutions easier for them to envision or implement than others (Criterion 2). Nor does the class address the manner in which the specific setting for any administrative work can influence how the technology in that setting is employed (Criterion 3), or the differential impact its use may have on the various actors in that setting (Criterion 4).

With the transition to Phase 3 and the infusion of technology throughout the curriculum as a medium of instruction more than as a tool for performing tasks, we do see more consideration of Question 2. When faculty teach via the Project Connect interactive video hookup or communicate with students via e-mail, one does see more discussion of how the characteristics of these technologies may influence the interactions they mediate.

On the other hand, the growth of Project Connect could also be seen to represent backsliding on Criterion 1: After Phase 2 introduced the treatment of technology as the means to some educational end, we have returned to a situation where a flashy new technology is to some extent employed simply because it has become available, rather than because it was carefully selected as the best way to meet some pressing educational need. (Whether one adopts such an interpretation would depend on why one believes this technology is funded and implemented. Does it have more to do with greater effectiveness of instruction or the promise of reducing expenditures on teaching staff as instruction is beamed to multiple locations simultaneously?)

Meanwhile Questions 3 and 4 continue to be neglected. One might ask if, for instance, regardless of why video-based distance learning technologies are originally implemented, such technologies are likely to result—in a context of intense pressure to contain costs and be accountable for a consistent level of educational outcomes—in efforts by educational institutions to wrest control of videotaped lessons from the individual instructors and reserve the right to rebroadcast the instruction indefinitely with no further compensation to the instructor.

Conclusions

If the experiences at UB are indicative of changes elsewhere, then what are the implications for the use of technology in administrator preparation? The chronology of technology use at UB supports two observations.

First, performance according to Criteria 1 and 2 has varied, illustrating the uneven evolution of educational thought, with each step forward being in some ways also a step back. Viewing the period as a whole, though, one can say the overall trend has been one of gradual progress with respect to the first two questions, despite occasional backsliding in one regard or another.

Second, and in marked contrast to the first point, the neglect of Criteria 3 and 4 has been quite constant throughout this period, indicating an area where additional attention might prove quite beneficial. To date, technology has primarily been treated as a source of increased efficiency, with little regard for the context in which it is employed or for the equity and fairness of its impact.

If the use of technology in reformed administrator preparation is to have a moral purpose, then we must ask (and prepare our students to ask), Who is using it? Why? And toward what ends? And perhaps more important, who will be helped, and how? Who will be harmed, and how? Herein lies an opportunity for genuine reflection and meaningful change.

Note

1. This experience was later reported in Jacobson, Westbrook, and Boyd (1994).

References

Apple, M. W. (1993). *Official knowledge: Democratic education in a conservative age.* New York: Routledge.

Bailey, G., & Lumley, D. (1994, Winter). Create a culture of technology. *Technology & Education,* supplement to *Electronic Learning Magazine,* p. 6.

Bluhm, H. P. (1987). *Administrative uses of computers in the schools.* Englewood Cliffs, NJ: Prentice Hall.

Carnegie Forum on Education and the Economy. (1986). *A nation prepared: Teachers for the 21st century.* Washington, DC: Author.

Carr, A. A., & Bromley, H. (1997, Winter/Spring). Technology and change: Pre-service teacher perceptions and agency. *Teaching Education, 8*(2), 15-22.

Crawford, R. (1994). *Computer-assisted crises* [Essay available at the International Philosophy Preprint Exchange at ftp://Phil-Preprints.L.Chiba-U.ac.jp/pub/preprints/Political_Phil/Crawford.Computer-assisted_Crises/tek_crit.txt]. (A condensed version of this essay appears in G. Gerbner, H. Mowlana, & H. I. Schiller [Eds.], *Invisible crises*, Boulder, CO: Westview, 1996.)

Fullan, M. (1993). *Change forces: Probing the depths of educational reform.* Bristol, PA: Falmer.

Gustafson, T. J. (1985). *Microcomputers and educational administration.* Englewood Cliffs, NJ: Prentice Hall.

Hodas, S. (1993). Technology refusal and the organizational culture of schools. *Education Policy Analysis Archives, 1*(10) [EPAA is a peer-reviewed journal distributed electronically. The Hodas article is available at http://seamonkey.ed.asu.edu/epaa/v1n10.html].

Holmes Group. (1986). *Tomorrow's teachers: A report of the Holmes Group.* East Lansing, MI: Author.

Jacobson, S. L., & Berne, R. (1993). *Reforming education: The emerging systemic approach.* Newbury Park, CA: Corwin.

Jacobson, S. L., & Brosnan, P. (1988). *Macintosh tutorial: Issues and applications of computers in educational administration.* Needham Heights, MA: Copy Right Press.

Jacobson, S. L., Westbrook, K., & Boyd, B. (1994, October). Developing a "virtual" community of international educators. Roundtable session at the annual meeting of the University Council for Educational Administration, Philadelphia.

Miller, H. (1988). *An administrator's manual for the use of microcomputers in the schools.* Englewood Cliffs, NJ: Prentice Hall.

National Commission on Excellence in Education. (1983). *A nation at risk: The imperative for educational reform.* Washington, DC: Government Printing Office.

Richards, C. (1989). *Microcomputer applications for strategic management in education.* New York: Longman.

Roszak, T. (1986). *The cult of information: The folklore of computers and the true art of thinking.* New York: Pantheon.

Sterling, B. (1993, May 10). [Speech delivered at the National Academy of Sciences, Convocation on Technology and Education, Washington, DC]. Published in *Computer Underground Digest, 5*(54, July 21, 1993) http://venus.soci.niu.edu/~cudigest/CUDS5/cud554.txt

Sutton, R. E. (1991, Winter). Equity and computers in the schools: A decade of research. *Review of Educational Research, 61*(4), 475-503.

Winner, L. (1986). *The whale and the reactor: A search for limits in an age of high technology.* Chicago: University of Chicago Press.

University Council for Educational Administration. (1987). *Leaders for America's schools.* Tempe, AZ: Author.

NINE

Preparing Teachers to Teach With Technology
THE COSTS AND BENEFITS OF DEVELOPING AN ELECTRONIC COMMUNITY OF LEARNERS

MICHAEL L. WAUGH

MARIANNE HANDLER

The purpose of this paper is to convey our perspectives regarding the costs and benefits of integrating instructional technologies in a specific instructional context in higher education. Because of varying institutional constraints, each of us is following a slightly different path in seeking to integrate computer-based instructional technologies in our teacher education programs. Yet, for both of us, the purpose of our efforts is to improve both the process and outcomes of teacher education. Although we have concentrated our efforts in this particu-

AUTHORS' NOTE: Some of the information conveyed in this chapter is based on work supported by the National Science Foundation under Grant No. RED-9253423. The government has certain rights in this material. Any opinions, findings, and conclusions or recommendations expressed in this material are those of the authors and do not necessarily reflect the views of the National Science Foundation.

lar area and on a limited number of computer-based instructional technologies, many of the problems we have encountered are similar if not identical to those faced by administrators and faculty at all levels of formal education and regardless of the specific instructional technology and instructional application of interest (U.S. Congress, Office of Technology Assessment [OTA], 1995).

There are many computer-based instructional technologies and many specific applications of them that may improve teaching and learning. Over the past several years our interests at the University of Illinois have focused on a particular technology—microcomputers linked together by digital data networks (often referred to as simply *electronic networking*)—and its potential for improving teacher education through developing on-line communities of higher education and K-12 students, faculty, and administrators who share common needs and goals and are willing to collaborate to promote learning. At National-Lewis University we have been working toward faculty use of integrating both computer-based activities and building better electronic networking opportunities.

From our perspective as teacher educators, engaging in electronic networking activities means to employ a combination of microcomputers, electronic networks, tools, and network use software in teacher education program coursework. The specific instructional purpose or application may vary widely across courses and programs, so we briefly describe two different approaches to illustrate some of the variations that are possible in teacher education. Likewise, the full range of possible instructional applications for electronic networks in K-12 is quite large, but we will not focus on these more than to briefly mention that our efforts in higher education are linked to those in K-12 as we both seek collaborations that will benefit all students involved. Furthermore, we recognize the commonality of resource and professional development needs that we all face (Executive Office of the President, 1997).

The Complex Nature of the Teacher Education Enterprise

Teacher education is a highly complex endeavor—an elaborate ecology—involving many interacting constituencies. It is the formal structure in which higher education faculty with expertise in several content areas and a specialized knowledge of schools, teaching strategies, and human psychology work to convey their knowledge in a systematic way to their students. The design of teacher education programs also involves faculty, students, and administrators from the precollege schools in numerous, significant ways as the college students progress through their formal studies and make the transition from student to teacher.

Over the years, numerous reports have criticized teacher education as being too conservative, too resilient to change (Carnegie Forum on Education and the Economy, 1986; Holmes Group, 1986). There is undoubtedly some truth to these claims, but it is equally true that the schools themselves change very slowly (Fullan & Stiegelbauer, 1991). A commonly held misconception is that if teacher education would change, then the schools would also quickly change. This is no more true than the converse that if schools would change then so would teacher education. An elaborate feedback mechanism is at work in which the "products" of the precollege schools become the "raw material" for a process in higher education that must in turn produce a "product" (new teachers) that meets with the approval of the state and local governance structures, that is, that fits their conception of the needs of the schools. Young people wishing to enter the teaching profession often find that they (a) understood "education" to be one thing (based on their lengthy, successful experiences) when they were precollege students, (b) were told that it should be a different thing when they were in college, and (c) felt compelled to choose between the two conceptions or adopt yet a third conception imposed on them by the requirements of the state. In addition, these initial conceptions are then shaped by the culture of the specific school in which the students finally practice their craft.

Much more could be said about this process, but the part that is most relevant to the purposes of this chapter is simply that if some change in "the system"—the way in which teachers are prepared to meet the needs of students—is considered to be desirable, then, in our view, it is extremely important that all elements of the system take actions that are consistent and mutually supportive of the other elements in the system. In the context of the acquisition and utilization of instructional technology, if we want computer-based instructional technologies to become a part of the culture of schooling, if we want students and teachers to use instructional technologies in appropriate and meaningful ways, then the precollege schools; higher education; teacher education; and local, state, and federal governments must collaborate to produce a clear set of expectations and appropriate resources and training (U.S. Congress, OTA, 1995). Although this view may seem obvious, it is still worth stating: Effecting systemic change is not a simple process, and the likelihood of achieving it is greatly enhanced through concerted effort by all stakeholders.

The Cost of Meaningful Change

In the paragraph that follows, we use the term *teacher* exclusively for the sake of consistency, but our comments are intended to apply to all teachers regardless of the level at which they teach.

Changes in social systems can be forced, bought, or cajoled, but if these changes are to persist, to become relatively permanent, then they must be accepted by the individual participant. One of the biggest obstacles to meaningful change in the field of education is the teacher's epistemology (Branson, 1993). By and large, teachers are incredibly competent and well-meaning individuals who are motivated by the sincere desire to provide their students with the very best educational experiences possible. Because of this, and because they themselves excelled in the traditional, formal process of education, many are hard to convince that there might be a better way to do "education." This idea has major implications for any teacher education activity, be it pre-service, in-service, or an extended series of experiences that might be called lifelong learning. Essentially, when it comes to the notion of change in either what they teach or how they teach it, many teachers naturally resist efforts to rethink their instructional strategies and style. It is possible and relatively easy to help teachers develop new skills and techniques, but ensuring that these skills are thoroughly integrated into their schema of what constitutes good teaching/what should be taught or into existing rigid curriculum sequences is quite another matter. In working with teachers, if change in content, teaching strategy, and style is the goal, then the teachers need adequate time to work with the new ideas, and they should be supported during this endeavor (U.S. Congress, OTA, 1995). For most teachers, their understanding of what it means to be a good teacher has developed over a long period of time. They cannot easily change these deeply held beliefs (Fullan & Stiegelbauer, 1991). It takes time, support, and a consistent message for the metamorphosis to occur.

In working with higher education faculty over the past several years and seeking to increase the utilization of instructional technology in teacher education, we have learned a great deal about the barriers that inhibit change in this setting. The system creates barriers that ensure conformity and limit creative invention. Some argue that this is essential, whereas others argue that these barriers are fatal to any initiatives that promote change. Here are some reactions that we have observed from higher education faculty over the years in response to our attempts to influence increased use of instructional technology in teacher education curricula: I can't spend any time working with that "computer stuff" because

- I must cover the content.
- It is too difficult to get access to the lab.
- No one is available to help me set up the equipment.
- My students do not have adequate access to the resources necessary to do the assignments that require technology.
- If I don't lecture, we are cheating them.
- I have a full-time job now. Where do I find the time to devote to this new stuff?
- If I become the expert, then I'll be asked to spend even more time helping others.
- If I invest my time in learning these skills instead of doing the things that the system rewards, I lose. What will it get me? Tenure? A pay raise?

Many, if not most, of these same attitudes are also voiced by some K-12 practitioners. There are many specific ways in which higher education and precollege education differ, but at the level of the individual teacher's attitudes and concerns, they seem far more similar than different.

Developing an Electronic Community of Learners: Two Different Approaches

The Teaching Teleapprenticeships Project at the University of Illinois at Urbana–Champaign

In 1992, the College of Education at the University of Illinois at Urbana–Champaign began a 3-year National Science Foundation-supported project, known as the Teaching Teleapprenticeships Project (Levin & Waugh, 1995; Levin, Waugh, Brown, & Clift, 1994), to examine the effects of infusing telecommunications technology into our teacher education programs. The main focus of the project was to provide support to the teacher education instructors in the college and to engage them in a process of reflecting on their coursework and looking for ways in which electronic mail, electronic conferencing, gopher, the World Wide Web, and newly emerging capabilities such as digital audio and videoconferencing could help them to improve their instruction.

The concept of teleapprenticeships was broadly defined so as not to exclude any viable initiatives proposed by faculty. The general patterns that evolved were activities in which the undergraduate teacher education students used e-mail to act as mentors to K-12 students, activities in which the undergraduates contacted experts in various fields and apprenticed under them for a specified period, and activities in which the undergraduate and graduate students acted as mediators in on-line projects that involved K-12 students and adult experts. In addition, the project developed a number of innovative frameworks for engaging the undergraduate teacher education students in activities that resulted in their creating useful gopher and web-based instructional resources.

A major aspect of the project involved the use of 72 Macintosh Powerbooks as tools for helping to keep the undergraduate students connected to their instructors and supervisors during their student teaching. Each Powerbook came equipped with a 14.4KB modem and a wide variety of software packages, such as *MacPPP* (for IP connection), *Eudora* (e-mail), *Netscape* (World Wide Web), *Newswatcher* (newsgroups), *Microsoft Works*, *KidPix*, and a variety of utility programs.

Throughout the project, the project staff provided training for the university instructors, the public school cooperating teachers, and the undergraduate and graduate students. Project staff also maintained the Powerbooks and provided assistance with telecommunications problems, software problems, and printing needs. In addition, they created and maintained the project gopher and web servers (http:// www.ed.uiuc.edu/tta/) and an e-mail server, FTP server, and an *Applesearch* server.

A very critical aspect of the Teaching Teleapprenticeships Project was ensuring that our professors and on-campus students would be able to join this electronic community as it evolved. We were fortunate in being able to begin to address this aspect of the project through a previous project begun in 1990. At that time, we were able to provide all of our faculty members with Macintosh computers and a network connection for electronic mail, printing, and file sharing. We were also able to begin the process of training our faculty and staff so that they would be prepared to join in these activities. To address the needs of our on-campus students, we established a computer laboratory with 28 Macintosh and 12 DOS/Windows-compatible computers. Furthermore, we equipped two teaching classrooms with computers, computer projection equipment, and Internet connections so that faculty could make presentations and display Internet resources to an entire class of students.

We undertook the Teaching Teleapprenticeships Project to study how faculty would utilize these communication tools and network-based resources to enhance their instruction and to effectively promote these changes. We wanted to learn what would be required to support their efforts to infuse technology into their curricula. We wanted to learn about the pros and cons of using this medium for instruction. In addition, it seemed obvious that the Internet would soon provide an incredible array of instructional resources for educators at all levels and so we wanted to ensure that teacher education graduates from Illinois possessed the basic knowledge and skills to be able to harness this emerging instructional resource. Furthermore, we wanted to create a human infrastructure consisting of university instructors, teaching assistants, graduate students, and undergraduate students who were sufficiently well equipped and skilled to be able to use this medium to engage in instructional interactions with others.

During the same period in which the Teaching Teleapprenticeships Project was being conducted, the faculty in the Department of Curriculum and Instruction were undergoing a major effort to redesign our undergraduate programs in the Elementary, Secondary, and Early Childhood areas. These redesign efforts were influenced by the project and these new instructional programs will incorporate a significant emphasis on content-appropriate instructional technologies. During the 1996-1997 academic year, the department pilot-tested the Technology Competencies Database, a web-based software application to assist faculty and students in addressing the ISTE/NCATE technology foundation standards. This database will enable us to coordinate the technology infusion efforts of faculty across these programs.

Grassroots Infusion at National–Louis University (NLU)

This section describes the efforts of one private, highly tuition dependent college of education, but it is likely to characterize the efforts at many institutions struggling to bring teacher education in line with 21st century educational efforts. Our College of Education, as is the case with many others, shares the resource "pie" with two other colleges the needs of which must also be considered. Some of the technology needs are common across colleges whereas others focus on the tools needed in the preparation of future teachers.

There is university-wide recognition of the need to create a stronger technological infrastructure if we are to appropriately integrate technology into our programs. The faculty of the College of Education at NLU are interested in making this effort but are often frustrated in their efforts. These frustrations include

- Lack of available computer labs
- Lack of open access computing sites that provide computers for student use in completing assignments
- Lack of support staff to set up the technology needed in instructional settings

We are trying in the ways that we can to begin providing technology experiences for our students. It is a challenge for us to move forward with limited resources. Thus far, we have been able to work within our existing institutional framework to take several significant steps toward success. To begin, we have one of the key ingredients to bringing about change: leadership and support at the administrative level. The dean of the College of Education provides whatever support she can, and this support has taken several forms. She has encouraged (within budget limits) the purchase of technology for faculty and student use.

Our dean also has provided opportunities for teacher education faculty to broaden their understanding and experiences with the instructional uses of technology. During the 1993–1994 school year, the Technology in Education faculty along with Elementary Education methods faculty who were integrating technology experiences into their classes planned a full day of training for the Elementary Education Department. There were large group discussions of where technology experiences fit into our curricula; there were hands-on sessions to try out instructional software; and there were demonstrations of model lessons for the attending faculty. The faculty were most interested in hearing an adjunct faculty member (a teacher in a local school district) demonstrate and discuss the ways in which the *Jasper Woodbury* videodisk series was used with students in her district. The day was opened by the dean and closed by the dean. In that closing segment, there was an opportunity to plan for the next steps. During the 1994-1995 school year, there were two additional opportunities for faculty development in this area.

During the following year, the college invited Dr. Jerry Willis (currently at Iowa State University) to meet with faculty in the Elementary Education and the Educational Leadership programs to consider the elements needed for technology integration to take place. Later that same year, Dr. Judi Harris, from the University of Texas at Austin, spent several days introducing Elementary Education faculty and adjunct faculty of the Technology in Education Department to the power of telecommunications for themselves and for students. The following year, a special section of a course on the World Wide Web was offered during the day for interested faculty and staffed by a member of the Technology in Education program. This ongoing administrative support has provided encouragement despite the serious lack of resources that continue to plague our efforts.

Our undergraduate pre-service students are required to take an introductory course in using instructional technology prior to their methods courses and student teaching. This assures faculty that students entering their classes have developed some competency with the computer and thus the faculty can deal with the application of the technology tools and not teach the students how to use the tools. Elementary Education faculty are developing plans to meet the ISTE foundation standards to help our graduates as they enter the marketplace.

Our institution is located in the Chicago area, and we offer our teacher education program at four campus locations. Each of these locations has a Windows and a Macintosh lab. Of the four Macintosh labs, two are well equipped to fully meet student needs. All of the Macintosh computers on one campus have been replaced this year. On three of the campuses, it is possible to use the Windows computers. On the fourth campus, the IBM-compatible computers are obsolete.

Limited institutional resources have been provided for academic technology needs. Our current system does not provide e-mail for all faculty, let alone each student in the teacher education program or other graduate education programs. Limited server capability does not allow for providing the needed number of IP addresses. The structure of the servers and the 56KB connection between the campuses and our Internet service provider do not allow us to make use of desktop conferencing opportunities.

NLU has four interactive video classrooms. There is need for increased professional development in how to best use this resource and how to develop instructional units. This links back to our Internet connections to provide the e-mail, listserv, and group conferencing needed to create real learning environments in this new format.

During the 1996-1997 academic year, institutional resources became available to upgrade three of the four Macintosh labs. The earlier models were then used to address three areas of need: (a) student walk-in labs of four to six computers were created, (b) computers were placed on carts to make them available for classroom instruction, and (c) additional faculty members were provided with computers on their desks. Administrative support was also provided for helping faculty to begin using the ISTE technology foundation standards in the redesign of their methods courses.

The higher administration at NLU is aware of the difficulties and of our need to move forward. Plans are currently being formulated to correct some of the frustrations faculty are facing. Private institutions do not have access to the state funding available to K-12 education as well as other state universities. Often, funding agencies exclude private colleges from submitting grant proposals. These resource issues impact on all of the components needed to implement a full-fledged, integrated technology environment for our students.

The Costs of Implementation

In the sections that follow, we discuss three critically important factors related to the costs of implementing an electronic community of learners: access and local resources, training, and support. To a very great extent, these factors are critical regardless of the specific instructional technology being considered. We discuss them in the context of developing an electronic community of learners, however.

There is a certain logic to discussing these factors in this order as one first needs to have *the necessary tools* to be able to do a thing, then must *know how* to do the thing, and subsequently must *be supported* in doing the thing. These factors are also critically interdependent and must be addressed simultaneously, or nearly so, in actual implementation (U.S. Congress, OTA, 1995). No technology initiative can succeed without addressing each of these three critical factors in an acceptable fashion. As with a three-legged stool, two legs are all but useless.

Universal Access/Local Resources

As it relates to this discussion, *universal access* may mean two distinctly different things, both of which are important. In the first sense, it means students and teachers should have access to adequate computers and equipment at their physical location so they can utilize the equipment for instructional purposes. In the second sense, universal access means that all students should be offered an equal opportunity to access the wide array of instructional resources that are becoming available via the Internet (in addition to those that are available through other distribution mechanisms and may be stored on a local server). That is, all students need the same kind of network access opportunities. In both senses, equity of access is critically important to achieve equity in *participation*, and this is as important for a free society as the very idea of universal education.

Technology should be used in *all schools* in ways that mirror its uses in our society. For this to happen, the hardware and software in schools must be kept current. The staff members in the schools must be trained in the major applications software (productivity tools) used in business and industry (the real world outside of schools) and in the network-based information access tools and instructional software delivery systems.

To expand on this notion of universal access, we describe the kinds of specific needs we have found to be essential for supporting teacher education programs on our campuses. Because we feel that these needs are shared across levels of education, we use the term *schools* to mean both precollege and higher education institutions. Likewise, the terms *teacher, faculty, classrooms,* and *students* should be interpreted as referring to entities across levels of education. Our perspective is that of professors in higher education, but our work in teacher education has shown us that these basic needs are common throughout the educational community.

Faculty Need Computers. Faculty must first *use* computers before we can expect them to use computers in their teaching. Until faculty are comfortable using computers, they will be reluctant to integrate them into their teaching. If we want tomorrow's teachers (citizens) to be computer-using teachers (citizens), then today's teachers must be appropriate role models. In *Schoolteacher,* Lortie (1975) reminds us that teachers teach as they were taught. Our teacher education programs do not give us extensive time to change all of the preconceptions about teaching that our teacher education students hold. Faculty need to have the best possible resources and support to both model the use of technology in instructional process and develop meaningful, technology-enhanced assignments such that their students can become comfortable thinking of technology as an essential instructional tool.

Students Need Computers. At a minimum, students need extensive access opportunities, and at best, they need the kind of unrestricted access that can feed creativity and provide them with opportunities to explore beyond the minimum expectations of their course requirements. Faculty should actually stimulate these kinds of independent research activities as these activities most closely mirror the independent lifelong learning activities that will be essential for success in the 21st century.

Schools Need Computer Laboratories. In many instances, a computer laboratory is the most efficient setting in which to accomplish group instruction. Demonstration is a valuable instructional strategy. Often, faculty need a facility where each student has access to a computer during the class. At other times, one computer for demonstration is sufficient and can provide additional opportunities for faculty to illustrate the ways in which they use technology as a tool for instruction. In this case, it would be a waste of a computer lab facility to fill it with students who did not need to use all of the computers as part of the class session. In fact, using a computer lab for demonstration or lecture would provide a negative model, an example of a misuse of instructional resources in a way that subtly reinforces such misuse.

Classrooms Need Computers. A phenomenon we have noticed often may be described as the Proximity Rule. The Proximity Rule is defined as, "The closer a resource can be located to the person who must use it, the more likely that person is to use it." In the context of teachers and technology, the closer the computers and network connections can be put to the teacher, the more likely it becomes that the teacher will use them. This proximity is necessary but not sufficient, however, to ensure that the resources will actually be used in instruction. We have often seen a situation in schools that sends precisely the wrong message to students. One or more computers are physically present in the classroom but are *never* used by the teacher. What message does this convey to the students about the value and importance of computers? Many of the students in these classrooms will be tomorrow's teachers. If they have years of similar experiences, how long will it take to change their views of how computers should be used in the classroom?

Although the previous discussion focused on hardware and access, the Proximity Rule also applies to professional development and support. If appropriate professional development (i.e., timing, quality, complexity) is not readily available, then faculty will not seek it out. If support is not readily available, then it will take very few equipment or software failures to discourage faculty and convince them that attempting to use technology in their classrooms is counterproductive.

To improve equipment access opportunities, classrooms need computer resources. Teacher education classrooms should be equipped with network connections and, ideally, with several computers that remain in the rooms. Teacher education programs should model the kinds of facilities that would be optimal for K-12, and the trend in recent years is the migration of computers out of computer labs and into classrooms. Likewise, the precollege schools should be equipped so that new teachers who have been trained to expect network connections and computers in their classrooms will have them and be able to put them to immediate instructional use.

Software. Along with computer laboratories and classroom computers, all schools need to ensure that adequate funds are made available to purchase new software and software upgrades. Software such as word processors, spreadsheets, databases, and telecommunications tools are changing rapidly, and the goal should be to keep pace with these changes to ensure that the software being used in the schools reflects the state-of-the-art. Although it may certainly be possible to use older software with students and help them attain basic ideas and skills, there is a subtle positive psychological influence associated with using tools that are the "latest and greatest," and this motivational boost can be very beneficial. Furthermore, if students are learning state-of-the-art skills and techniques, they *instantly* become valuable human resources in the "real world" outside of school.

Although some faculty might resist changing hardware and software merely for the sake of change, often the newer products are significantly more capable than their predecessors and so the upgrade process is desirable for many reasons. But faculty should not be forced into making changes whenever new software becomes available; rather, these new materials should be made available and faculty should be offered encouragement and support to help them make the change. Just as there is a subtle positive psychological influence associated with using the latest software products, there is a subtle negative one associated with clinging to an older, obsolete product. Given the opportunity to improve their situation, faculty (just like everyone else) will eventually choose to do so. We need to give teachers the opportunity and constant encouragement to "change for the better."

Student Ownership. Another access issue that we are grappling with in higher education involves the potential for student ownership of computers as opposed to institutional ownership of the computers that students are required to use. Each of these options has desirable attributes and illustrates important issues for resource planning and management.

It can be very beneficial for teacher education students to own their own computers. This provides students with a powerful learning resource and enables them to gain a much wider range of experiences with computers during their studies. Ownership maximizes the potential predicted by the Proximity Rule. It also permits students to acquire software for their computers through institutional licensing agreements. This can enable the student to benefit in one of two ways. If the licensing agreements permit the students to keep the software when they leave the university, then they have software to use when they begin teaching. If the licensing agreements do not permit the students to keep the software they were able to use during their training, then the students have benefited by deferring the purchase of the software and by developing skills with the particular software packages so that they will be more informed consumers when they make software purchases of their own.

Student ownership of computers helps address a critical problem for higher education institutions, and this issue may also be of importance to K-12 schools at some point in the future. When students purchase their own equipment, they take it with them through the system and the organization is not left with large quantities of old equipment to repair, maintain, or replace. This relieves the institution from having to purchase new equipment in sufficient quantities to address the needs of all the students, all of the time. Instead, the institution need only to be concerned with meeting the needs of some of its students, some of the time through computer laboratories and classroom clusters and the purchase of special hardware or software for specialized needs.

Student ownership also presents problems for an educational institution. If more students own their own computers, then the overall number of computers and equipment configurations (hardware as well as software) increases dramatically and this, in turn, places a significantly greater burden on the institution (if the institution desires that these computer resources be used to enhance instruction) to provide increased technical and instructional support and to accommodate dramatically increased demand for access to the network through the institution. Also, at the present time, there are two major "families" of personal computer hardware platforms available in the marketplace. Both hardware platforms are used in business and education. Because of this, and because faculty members might require that students work with particular software applications unique to a particular computer, educational institutions should provide computer labs that contain both of these platforms, since it is unlikely that students will own both types of hardware.

Regardless of the perceived value of student ownership, the number of university students who own computers or have access to computers where they live is increasing steadily, and institutions should be aware of this fact. A survey of the elementary education majors at the University of Illinois in 1995 revealed that approximately 40% of those students owned a personal computer at that time (although the university does not require its undergraduates to purchase computers). Other units at the university also report similar percentages regarding their undergraduate student population (personal communication, discussions of the Educational Technologies Board at the University of Illinois, November 1995).

Professional Development

Despite being listed second among the critical factors, professional development is perhaps the most important factor of the three. Without training and experience, too many computers and network connections remain unused. Professional development is also the most critical of the three factors because it takes the greatest amount of time to achieve. In our opinion, training should be started as early as possible to begin the lengthy process of changing attitudes and philosophies.

Synchronizing Supply and Demand. Two examples illustrate the importance of beginning professional development activities as soon as possible. The first example is straightforward. In the Teaching Teleapprenticeships Project, we were able to make significant progress during the term of the grant because we had previously taken steps to provide faculty with resources and training so they could become comfortable with using networks and e-mail *before* they were challenged to look for ways to employ these resources in their teaching. Throughout the term of the grant, we continued to offer professional development opportunities to faculty, but the majority of these activities were focused on curricular integration efforts rather than hardware and software basics.

A second example concerns the nature of Internet connections and a logical strategy for planning and implementing Internet access in schools. This example is slightly more complicated, but it serves to illustrate the complexity and interrelatedness of the professional development and universal access/local resources factors.

There are two basic ways that individuals and organizations can connect to the Internet. One way is through common telephone lines and the second method is through a high-speed, dedicated data circuit. These methods provide identical services, except that using common telephone lines is considerably slower and the speed of the connection is an important element in determining the success of instructional application. In addition, telephone lines typically cost less than a high-speed, dedicated data circuit, and they serve single users as opposed to groups, such as schools or districts.

Although high-speed, dedicated data circuit access is highly desirable and should be the ultimate long-term goal, should implementation of Internet access be delayed until a school or district can afford the higher cost of such service? Should professional development opportunities be delayed until access is available? Every organization desiring to connect to the Internet must answer these questions before spending funds to acquire the hardware to connect their constituents. The problem is more easily understood if it is rephrased. Should one hold off learning to drive until one has sufficient funds to buy a Cadillac? Assuming a critical need for transportation exists, the answer is clearly no. Learning to drive is a basic skill that can be acquired in any vehicle. This idea applies identically to the Internet access problem. Provide faculty with the best access that is affordable *immediately* so they can be trained and *begin* to develop the experience they will need to integrate Internet-based resources into their instruction.

Once it becomes possible to improve the means of Internet access (to buy a Cadillac), teachers will be able to immediately appreciate and utilize the enhanced capability. Network access and classroom application are quite different issues, and although the former can enhance the latter, the latter need not by stymied by limitations in the former. Another way of saying this is that some access is better than none, and many very valuable instructional applications can be accomplished with a less than optimal network connection (You can get there in a Yugo).

From the standpoint of budget managers, this is a highly desirable situation because it means that the increased expenditure of funds needed to improve the school's connection will be for a resource that can then immediately be put to use. It is pointless to pay for a resource that faculty are not yet ready or able to use. But budget managers should also be aware that it is highly counterproductive to prepare faculty to use a technology that is unavailable. Once faculty are trained to use the Internet in their courses, their demands may increase faster than the capacity of the institution's network to respond. When this happens, instructional efforts are thwarted, development efforts are delayed, attitudes suffer, and initiatives stall. Indeed, this one example clearly illustrates one of the major points we wish to make in this chapter, the need for synchronized effort to ensure success. *Supply and demand must be grown at the same rate to ensure maximum benefit from the investment.*

There is a potential negative side to initially providing less than optimal network access. This negative is the cost of the initial access system followed by the additional cost associated with changing over to the new access system. These costs are very real but should be compared with the long-term costs of delaying initial professional development, bearing in mind that a significant amount of time must be allowed for teachers to develop sufficient experience to become effective in using the new technology. In the context of electronic network access, the initial access method is likely to be through standard telephone lines, and the subsequent, enhanced access will be via a high-speed, dedicated data circuit. The good news is that these two methods are somewhat complementary, and so the initial cost for equipment to support the smaller group of "pioneers" may still provide a viable, though more limited, service to the organization as a whole after the upgrade to access via a high-speed data circuit.

In determining whether or not to initially invest in a dial-up access system, the critical issue is that of expansion, that is, expanding access to the larger group. It will be far more desirable and less expensive to provide high-speed data circuit access to the larger group than to expand a dial-up method to accommodate the needs of all members of the organization. Careful consideration must be given to the overall size of the group to be serviced and the timing of the changeover to maximize benefits and minimize costs.

Rates of Change. A fundamental principle of nature is that change is the only constant. Professional development should be planned with this notion in mind. Professional development should be a continuous process. Hardware and software are constantly changing, and through our experiences with using technology with students we are constantly learning about what works and doesn't work. Technology training should be organized to provide the proper instruction precisely when it is needed and to permit extended support. This is something that is ideally accomplished through involvement in an electronic community of learners. More experienced members of the group will be able to help those who are less experienced, and as the less experienced develop expertise they, in turn, will be able to help others. We call this highly individualized learning experience "just-in-time" learning. Because in the area of technology the pace of change is swift, the learning must be timely and continuous.

With due respect to music educators everywhere, we would like to describe a simple scenario that illustrates the importance of providing faculty members with more time and support for learning how to use instructional technologies. Some faculty development workshops in technology resemble one type of introductory instruction in music—the type that concentrates on vocabulary, definitions, and so on. With this approach, the student learns about the notes and the staff and the clefs. Later, the student learns to play an instrument and spends many hours perfecting these skills. After several years of study and practice, the student may learn to write music and compose symphonies.

These technology workshops for faculty begin in a similar fashion. The workshops focus on terms and concepts and even some procedures for using the hardware and software. But this is where the similarity ends. Having completed the technology workshop, faculty are expected to use the technology in their instruction. They are expected to be able to immediately innovate, integrate, and synthesize in the new medium. In essence, with minimal exposure to the technology and little or no practice time, faculty are expected to be able to instantly compose "technology symphonies."

These expectations are unrealistic, and staff development efforts that are conceptualized as one-shot workshops rather than a series of continuing experiences will only succeed in producing faculty who may be able to recall the name of yesterday's "truly revolutionary" instructional technology.

Support

Support is very critical to the overall success of any technology initiative. Two major types of support are technical and curricular support. Technical support focuses on the mechanics of the process, that is, the basic skills of what and how in regard to using the technology. Curricular support addresses the purpose of using technology, that is, the why. Despite the fact that faculty are content experts, in the sense of determining what should be presented, they very often need assistance in learning what the new technologies can do for them and how these new capabilities should be integrated into instructional strategies that will promote learning in their content domains.

As with the Proximity Rule regarding material resources described earlier, it is highly desirable to place content and technical support resources as near as possible to those who need them. We also have found that an important source of both technical and content support is the extended group of colleagues that make up the electronic community of learners, that is, the group of acquaintances and associates that they will encounter on the Internet.

Although recent software improvements and hardware developments make it seem as if technical support is becoming less important, this is merely an illusion. Electronic networking is a highly complex process. It is unreasonable to expect that anyone but a trained professional will be able to devote the time required to establishing and maintaining local and wide area networks. The work that is necessary behind the scenes is crucial to the success of any instructional effort that will involve resources from the electronic community of learners. Once teachers and students begin to make significant use of computers and networks, it is extremely important to ensure that the resource is operational and available at all times. Someone needs to be responsible for this on a full-time basis.

We have witnessed the problems that can occur when faculty and students invest significant effort in using electronic networks and then, through no fault of their own, the network goes down or the hardware crashes. It takes very few events of this type to convince teachers that this technology is not yet ready to be used in their classrooms. And it takes much more than an additional training session to rebuild the confidence that is necessary for teachers to allocate a valuable portion of their limited instructional time to this new instructional resource.

Funding Inconsistencies

In this chapter, we argue that with regard to instruction technology all levels of education face similar needs (Executive Office of the President, 1997). Further, we argue that it is extremely important for teacher preparation programs and the K-12 schools to coordinate their efforts so that tomorrow's teachers will understand the value and importance of using technology to enhance student learning. We must now acknowledge the fact that the formal funding mechanisms for higher education institutions and the public schools vary dramatically. Furthermore, federal and private funding opportunities are also highly variable (Executive Office of the President, 1997). Given these significant inconsistencies in funding opportunities, how is it possible to ensure that instructional technology will be integrated across all levels of formal education in a reasonably cohesive manner?

Although it is tempting to imagine that all institutions of higher education have significant funds to allocate for technology whereas school districts are chronically underfunded, this scenario is a fiction. The reality is far more complicated. In some cases, school districts are richly endowed and colleges (particularly colleges of education) are poorly endowed. In other cases, even the most poorly endowed college is rich in comparison to precollege school districts. In this chapter, we briefly described the situations that exist with regard to the instructional technology available for use in teacher education at a large public university and a private university, and the differences are evident. In the next section, we describe a well-funded public school district, and it should be obvious that its particular situation is well above the norm.

At the present time, higher education institutions and the precollege schools are not synchronized in the technological environments and experiences that they can offer their students, and the highly variable nature of governmental and private funding mechanisms has undoubtedly contributed to this state of affairs. In our view, inconsistent funding mechanisms are a significant impediment to effective collaboration between higher education and the K-12 schools, and this is detrimental to the development of technology-using teachers.

A Successful Resource Allocation Framework:
The Rule of Thirds

In this section, we briefly describe a resource allocation framework employed by Springfield School District 186, an urban, K-12 school district in central Illinois that has been highly successful in its efforts to acquire instructional technologies and utilize them in its instructional programs. District 186 employs approximately 1,000 teachers to serve the needs of approximately 15,000 students.

In 1984, Mike Holinga, Director of Technology for Springfield, (http://www.springfield.k12.il.us), developed the Rule of Thirds as a heuristic to guide the expenditure of the district's limited technology funds (personal communication, January 1993). As you will see, the general concept is not particularly revolutionary. At least one governmental report describes the need for a somewhat similar approach (Radlick, cited in U.S. Congress, OTA, 1995). And many of us have encountered funding schemes with similar characteristics. Springfield, however, has been adhering to this rule for approximately 12 years and demonstrates that consistent application of such a framework over time can create and sustain a technology-rich learning environment. Whether the proportions in the framework can or should be varied to adapt it for other organizational contexts is unclear, but this framework continues to be successful for Springfield.

The Rule of Thirds helps to quantify two of the issues described in this chapter. First, the rule provides a strategy for allocating funds for implementing technology in schools. Second, it provides a strategy for dealing with recurring equipment acquisition and maintenance costs.

Springfield's technology budget is composed of funds from two primary sources. Approximately 1% of the district's total annual operating budget is allocated to technology each year. In addition, Springfield receives funds for technology from several grant awards. In 1996, the total technology budget for the district was approximately $1.25 million. Holinga allocates these funds by spending approximately one-third on hardware purchases and repair, one-third on software and telecommunications charges, and one-third on training and support personnel. This framework for allocating funds has enabled the district to (a) create a wide area network linking all of its school buildings to the Internet through a 1.5 Mbit/sec dedicated data circuit; (b) provide dial-up access (for e-mail) for 1,500 district staff; and (c) establish a significant staff development program that provides initial training and follow-up support for district teachers. Springfield is widely known throughout Illinois as a leader in integrating technology into its elementary curricula, and it is now working to expand its program to the middle and high schools in the district.

In addition to providing a more comprehensive approach to the integration of technology into curricula, the Rule of Thirds also provides a reasonable strategy to be employed in addressing the recurring equipment acquisition issue. The purchase of computer equipment should not be viewed as a onetime event in which all of the organization's needs are met but rather as an ongoing process in which new equipment is purchased every year to augment or replace obsolete equipment in service (Firestone & Corbett, 1988). Given the short life span of computer technology (the relatively short period of time that the equipment is capable of running the newest and most advanced software), approximately one-third of an organization's equipment should be scheduled for initial acquisition or replacement in a given year; thus a complete turnover of equipment could be accomplished every 3 years. Equipment acquisitions are always problematic for school districts and institutions of higher education, and although this rule does not minimize the difficulty in obtaining the necessary funds to make such purchases, it provides a framework to guide the expenditure of funds. In this context, whether the actual implementation of the heuristic is as a rule of thirds, fourths, or fifths will depend on the needs of the organization and its ability to generate funds on a recurring basis; the critical aspect of the rule is the recognition of the recurring nature of this expenditure.

By definition, this strategy will only please a fraction of the people in an organization at any one time, but it will ensure that no one is forced to work for very long with obsolete equipment. It has the added advantage that it maps nicely onto the reality that not everyone in a college of education or a school system—or in any organization—is ready to begin using technology at the same time. Because of this, equipment acquisition and initial training can be staggered to better meet both the organization's and the individuals' needs.

Summary

In this chapter, we have argued that both higher education (especially teacher education) and precollege education have similar technology needs and that both should be acquiring and utilizing instructional technologies at the same rate in order for the two enterprises to work together most effectively in developing technology-using teachers or faculty. Although interinstitutional cooperation can be difficult for a variety of reasons, it is made much more difficult in this instance by the fact that the mechanisms for funding higher education and precollege education are considerably different (U.S. Congress, OTA, 1995).

Despite these differences, both educational enterprises have a need for guidance in planning and the expenditure of limited funds. The Rule of Thirds has demonstrated its value as a heuristic for helping to focus attention on the fact that a successful technology initiative must simultaneously address several critically important factors. Furthermore, the Rule of Thirds helps to illustrate the need for a recurring allocation of funds for technology, and it provides a framework for the application of those funds to help ensure that organizations avoid the problems associated with (a) spending funds before members are able to make use of what is purchased, and (b) the periodic mass purchase/mass obsolescence cycle.

References

Branson, R. K. (1993). Alternative models of schooling: Budget reallocation and policy change. *International Journal of Educational Research, 19*(2), 145-156.

Carnegie Forum on Education and the Economy. (1986). *A nation prepared: Teachers for the 21st century.* Washington, DC: Author.

Executive Office of the President. (1997, March). *Report to the president on the use of technology to strengthen K-12 education in the United States.* Washington, DC: Government Printing Office.

Firestone, W., & Corbett, H. (1988). Planned organizational change. In N. J. Boyan (Ed.), *Handbook of research on educational administration* (pp. 321-340). New York: Longman.

Fullan, M., & Stiegelbauer, S. (1991). *The new meaning of educational change* (2nd ed.). London: Cassell Educational Limited.

Holmes Group. (1986). *Tomorrow's teachers: A report of the Holmes Group.* East Lansing, MI: Author.

Levin, J. A., & Waugh, M. L. (1995). A college of education of tomorrow: A process for facing the challenge of technology today. *Journal of Computing in Teacher Education, 11*(4), 19-22.

Levin, J. A., Waugh, M. L., Brown, D., & Clift, R. (1994). Teaching teleapprenticeships: A new organizational framework for improving teacher education using electronic networks. *Journal of Machine-Mediated Learning, 4*(2-3), 149-161.

Lortie, D. C. (1975). *Schoolteacher: A sociological study.* Chicago: University of Chicago Press.

U. S. Congress, Office of Technology Assessment. (1995, April). *Teachers and technology: Making the connection* (OTA-EHR-616). Washington, DC: Government Printing Office.

Images of the Future
CRYSTAL BALL GAZING AND TECHNOLOGY POLICY MAKING

JAMES F. ANGEVINE

"If you do not think about the future, you cannot have one."

—JOHN GALSWORTHY (KNORDLUN, 1995)

As you will soon discover, this is *not* a scholarly research paper; nor is it an intellectual treatise. Rather, it is the simple musings of one who is intrigued by the fits and starts of change and advances in the various technologies.

It seems that it's the little things, the subtle indicators, that begin to get one thinking about what the world will be like in a few years and how one will fit in that new and different world. While sitting in the doctor's office a few months ago, I was perusing some *PC World* magazines, and an article caught my attention. This article dealt with Bill Gates's involvement in a *very* expensive telecommunications project. The goal of the project is to deploy a number of low-orbit (240 miles) communication satellites that would be used to relay information packets between clients and servers. Current satellites, which orbit at much higher levels, cannot be used for Internet traffic due to the time lag created by the great distance up and back that a packet would have to travel. This problem is virtually nonexistent when transmitting to the low-orbit satellites, which would provide unfettered access from anywhere in the world. Think of it! You're backpacking in the mountains of Tibet. You can pop open your laptop. Attach the little satellite dish. Fire it up, and you're on-line (or on space, since there are no "lines"). I know. Big deal!! But what are the ramifications of this simple (but very expensive) venture? As voice and video transmission through the Internet become more sophisticated, will we need phone lines or cable connections? What will happen to these industries? In that environment, where will "work" be done? Probably not in the same locations that we associate with "work" now. Offices, corporate headquarters, government centers, college campuses, and so on may become optional places to do "work."

At the risk of rehashing some old saws, the human race has undergone a number of significant and distinct changes over time. We started out as hunters and foragers living a nomadic tribal existence. Life was difficult and uncertain. Individuals banded together in tribes for protection, the rearing of young, the production of necessary utensils requiring special skills, rudimentary medicine, and so on. A loosely structured community that allowed each member to utilize his or her unique skills and abilities to support the varying needs of the community and ensure the passing on of needed skills and knowledge to the young in order to perpetuate the "life" of the community. Last, but not least, the community gave the members a sense of belonging to "something."

This model persisted for many thousands of years but was eventually shattered when we started learning how to plant seeds and not to kill everything with four legs. These "discoveries" heralded the onset of the agrarian period. Although farming and raising livestock became the primary means of supporting the family, the need for a community continued for many of the same reasons as in the foraging/tribal era. We still needed to rely on the expertise of others to provide specialized goods and services and to provide for the education and enculturation of the young. Notwithstanding the occasional crusade, scourge, colonization, New World discovery, and other relatively insignificant events (to the majority of the population), we planted the crops, tended the animals, bought and sold our products, and raised our kids for more than 3,000 years.

Then, we discovered that burning stuff (aside from ruining the air) could provide a way to transform dead plants and dinosaurs into energy to run machines to do "work." Bingo! The Industrial Age springs forth as the dominant cultural model. New industries begin to flourish. Lots of us move to where the jobs are. Cities begin to grow, and small farms are bought out by large food production industries. We all become dependent on the "system" (manufactured goods, food production, medicine, schooling/enculturation, transportation, entertainment, etc.) for our basic life-sustaining needs. Additionally, we develop a whole new set of "perceived needs," created through effective advertising and pushing the "keeping up with the Joneses" mystique. Schools become modeled after factories to train our youth for the industrial jobs (only a few for real thinking). Who cares about the dropout rate? We need these kids in the factories.

Communication systems develop to inform, educate, and market to the masses. Government at all levels increases in size and the scope of influence to support infrastructure development, using tax resources and regulatory functions to provide order in nearly all facets of the growing "centralized" community. Wealth and power become more concentrated in fewer hands in many nations. International marketing grows. We have wars, conflicts, police actions, and a growing military/industrial complex. In 1945, the Electronic Numerical Integrator and Computer (ENIAC) ushers in the Information Age (University of Pennsylvania, 1997). We beat the dickens out of our environment. Fear strikes, and we start to clean up the environment at great cost. Workers start making bigger wages. Whoops! Making stuff now costs too much. Let's just close down the factories and get the basic stuff manufactured in other countries where the wages are low. That'll save our profit margin. The current dominant cultural model begins to teeter.

The ongoing tension created by the way the victors divided the spoils of World War II (the Cold War) permeates the psyche of both the industrialized and the not-so-industrialized world. Serious questions about what happens after the "big one" caused a pervasive fear among the possible survivors. How will communications, knowledge, "commerce" (such as it might be) be maintained at some level after the expected cataclysm? Maybe a system of interconnected computer networks that could all "speak" the same language would do the trick? In 1969, we started to invest some serious dollars into this little endeavor (Public Broadcasting System, no date). The gestation period of the Internet begins. Soon, the DOS operating system strikes, and the masses start to compute. The network of networks becomes accessible to the nontechnical, nonspecialized consumers of information. We begin to talk among ourselves, share information, entertain ourselves, educate ourselves, and interact with one another on a one-to-one basis unfettered by geography or nationality. Planting seeds to feed ourselves, burning fossil fuels to power machines to do work, global communication and information access were, and are, harbingers of cultural upheaval and evolution.

Change is inevitable, but the speed with which changes are occurring now is unnerving. No! It is downright scary. We were hunters and gatherers for thousands of years. There was lots of time for ongoing correction of flaws in the cultural model. As planters of crops and tenders of animals, we had about 3,000 years to pick ourselves up after each cultural stumble. But 25 years from ENIAC to Internet doesn't allow for a whole lot of recovery time for mistakes in how we are structured to operate in our "expanded" global community.

What might a community "network" in a network of community "networks" look like? Are our institutions (i.e., universities, governments, social services, schools, etc.) prepared or preparing for the changes to come? Let's take a look at what a "node"—community "network" might look like in a network of community "networks."

Like all communities, the community "network" will be grounded on some economic base(s). The hunter/gatherer community was based on the herds and whatever edible vegetation was in the vicinity. As seasons changed and herds migrated, the community moved along with its economic base. Later, crops and livestock served as the primary economic base, and manufacturing, commerce, finance, government, education, and their associated institutions (buildings) served as the economic base for existing densely populated communities.

The speed of the technological development from ENIAC to Internet happened in the blink of an eye (in terms of previous cultural metamorphoses). This rate of change makes predicting, with any sense of certainty, the economic base(s) of the communities of the future a real problem. Speed tends to fuzz up the old crystal ball. Perhaps a look at *how* we might work in the future may provide some insight as to *what* we might be doing. Work will be decentralized, generally not dependent on structures or bound by geography. Industries will be specialized and flexible to accommodate the new (but short-lived) products demanded by an increasingly aware consumer in the global marketplace (a cybermarketeer). Education and training will become essential elements of "doing business." There will be a constant need to train and retrain ourselves as products, markets, services, and "new ways" of providing them expand. Given these "new" ways of doing business, here's one person's view of what a community "network" node might look like.

Imagine your favorite local "mini" mall. Empty all of the stores. You now have an easily accessible and flexible shell that is surrounded by a loosely tied group of individuals and families bound by some form of common or complementary economic base(s). The folks in the community don't need to *go* to work, but they do need to buy commodities, have medical checkups, be entertained, take care of their elderly parents, vote, get access to government (at all levels) programs and services, get items repaired, access the office center of the local economic base(s), converse, and so on. Now, let's fill up the remaining empty stores in the mall with what the community might need.

One-Stop Shopping. Most major goods and appliances will be ordered from regional warehouses via the Internet or a dedicated shopping intranet. Relatively small quantities of perishable foods, canned goods, and so on will be on the shelves in the Buy Stuff Mart, but most orders will probably be done in advance through the community "network" intranet. Items will then be packaged and available for customer pickup.

The Medical Center. The medical facility will be staffed by highly skilled medical technicians and nurses. They'll have access to a variety of telemedicine technologies for diagnosis and tissue and blood sample testing and constant access to specialized physicians worldwide. Emergency medical technicians will also be on staff to deal with critical incidents and trauma prior to medevac transport to regional full-service facilities.

The Community Center. Aside from cafes, coffee shops, and other types of watering holes, the community center would also provide for performing arts, sports events, lectures, and so on, which would be available via videocast and, on occasion, live performances. An elderly care center would be available, both full- and part-time, for the older members of the community. The center would be easily accessible to the Buy Stuff Mart, entertainment, the conservatory, and meeting places to encourage volunteerism and to ensure lots of interpersonal contact with the other adults in the community as well as the kids.

The Government Center. The government center would serve as a one-stop shop for all federal, state, and local government services, allowing the application for and issuance of licenses, access to financial and medical assistance programs, information on the status of governmental programs and legislative activities (if you don't want to bother getting the information at home on the Internet), contract and grant opportunities and applications, tax payments (some things are always a constant), and so on. There might even be a conservatory with flowers, benches, fountains, and so on enclosed and open for use all year. This would be a place where folks could talk, walk, relax, be alone, be together, meditate, have fun with the kids, and the like.

Schools. The school would be very different from the current school concept. It would be more of a learning center that orchestrates rather than teaches (in the traditional sense). But, more about schools later.

Well, there you have it. One person's somewhat cursory view of what features a future community "network" might contain.

Now, what would the school in this community "network" look like? What would the staff be like? How would instruction be delivered? Would classes be (as now) primarily self-contained? How would technology be used? How would "programs" be designed? How would the progress of children be assessed?

The school as a learning center would have a lead teacher (or teachers depending on the number of kids in the community) who would be highly knowledgeable in learning theory and its practical application in instruction—in other words, a good coach who knows the skills and abilities of each player and is able to provide the proper experiences and incentives at the right time to move individuals to be the best they can be. The lead teacher, working with the parents, the youngster, other school staff (both professional and skill oriented), and appropriate community resources, would be responsible for over-seeing the development of an individualized education plan (IEP— what an original term!) for each youngster. The plan would be developed after a period of assessment to determine the base knowledge level that the youngster brings to the learning experience and the learning style(s) of the individual.

A curriculum, specific to the youngster, would be developed deline-ating a sequential array of skills to be mastered in each content area with clear benchmarks that would be measurable and observable. The lead teacher would then identify other school staff and community members to work with the student based on the needs of the young-ster and the objectives of his or her unique program. These assign-ments would be crafted to ensure that the skills, personality, and teaching style of individuals working with a youngster are comple-mentary to the needs and learning style of the individual child. The lead teacher would be responsible for overseeing and interpreting the continual assessment of the youngster's progress and modifying the instructional plan to make sure that *all* of the elements of the IEP are achieved.

Actual "teaching" would be accomplished through a variety of means in a variety of settings. Emphases would be placed on individ-ual and group projects that would draw simultaneously on several content areas. With global access to information, youngsters (with good coaching) could use their parents, community resources, com-munity members, community businesses, school staff, virtual univer-sity programs, the libraries of the world, fellow students, peers from all nations, and more. The possibilities are endless.

Are we doing enough, learning enough, testing enough now to get it "right" when the time comes? Because that time is coming soon . . .

"Chance favors only the prepared mind."

—LOUIS PASTEUR (KNORDLUN, 1995)

Change in our current cultural model will occur. Will the shape of the new model be the result of actions based on informed policy deliberation and decision, or will it be shaped by reactions to an immediate political need, a threat, or self-interest? That responsibility rests with each of us who work in the areas of research, education, and policy.

But are we aggressively pursuing this responsibility? Let me ask a few questions.

- Are we actively seeking to create (through grants or contracts) opportunities to test alternative educational or community models that can be thoroughly analyzed in a somewhat controlled environment? If not, how can we hope to inform the inevitable policy decisions that must be made?
- Are schools of education in a position (philosophically, technologically, and attitudinally) to effectively train the educational specialist (the lead teacher) of the future?
- Are we looking at our use of infrastructure resources internationally? Should we try to be in a position to advise decision makers on the best use of these funds? Are large, hard-wired, multistory school buildings (the old factory model) a viable piece of a future model, or are we condoning dinosaurs?
- Are our educational institutions gearing up for the rapid response demands of training and retraining individuals to maintain economic viability in a fast-paced, technology-driven economy?
- Are we forging international linkages among professionals in anticipation of the need for a global learning network accessible by practitioners and children in the future?

The list could go on, but the real question is

- *Are our institutions and professional organizations actively and consciously pursuing an agenda that anticipates a future with significantly different sets of operating rules and structures? Or are we simply analyzing and reaffirming what is and what has been?*

The crystal ball is beginning to clear just a bit.

"We are on the verge of a revolution that is just as profound as the change in the economy that came with the industrial revolution. Soon electronic networks will allow people to transcend the barriers of time and distance and take advantage of global markets and business opportunities not even imaginable today, opening up a new world of economic possibility and progress."

—VICE PRESIDENT ALBERT GORE, JR.
(INFORMATION INFRASTRUCTURE TASK FORCE, 1997)

References

Information Infrastructure Task Force. (1997, July). [On-line]. Available: http://www.iitf.nist.gov/eleccomm/ecomm.htm

Knordlun, K. (1995, August). [On-line]. Available: http://beam.helsinki.fi/~knordlun/misc/quotations.txt

Public Broadcasting System. (no date). [On-line]. Available: http://www.pbs.org/internet/timeline/

University of Pennsylvania. (1997, April). [On-line]. Available: http://homepage.seas.upenn.edu/~museum/overview.html

ELEVEN

Bits 'n' Bytes
EPILOGUE AND PROLOGUE

KATHLEEN C. WESTBROOK

And so we come to the end of the 1997 Yearbook of the American Education Finance Association. What have we learned? Where have we been, and where are we going? How can we help our future administrators, our state legislatures, and our local school boards? What raison d'être do we have for caring about educational technology in local, state, or international communities? Or is the technology too complex, intangible, and "loosey-goosey" to even bother? Are there just too many variables, too few answers, and does anybody even care?

DATELINE: Fairfax County, Virginia
BUY IT AND THEY WILL LEARN

More than 2800 pieces of classroom computers, printers, or terminals are broken or neglected in Fairfax County (VA) public schools. A school official says: "The focus of attention was on buying the equipment, and the support of that equipment was not taken into account. It was assumed the current support

184

systems would be able to handle things and that has not proven to be the case." The school board's budget panel chief says the board's decision not to hire additional technicians for this fiscal year was influenced by its budget policy to hire administrators only when absolutely necessary. (*Washington Times*, 1997)

If you don't notice what's wrong with the above paragraph, you have probably not read any of the chapters in this volume or are one of the many who believe anyone not in a classroom *must*, by definition, be an administrator. Says whom? Well, in many places, just about everyone. In Illinois, for example, you can be a teacher or an administrator or an Allied School Professional (social workers, psychologists, school nurses, secretaries, etc.), but you cannot simply be a highly skilled individual needed to maintain our ever-growing inventory of technology equipment.

Why *Not* Administrator?

First, most school administrators are not "techies." They have spent the past 5 to 10 years of their professional lives learning and preparing to manage curriculum, staffing, budgets, teachers, and students. Their training did not include keeping computers, local area networks, wide area networks, "clouds," cable connections, routers, modems, client-server software, intranets, the Internet, and security firewalls installed, let alone working.

Second, their focus, by design, is on the needs of students and curriculum—and it should be. Although the myriad of noneducational tasks administrators are asked to perform increases each year, to this should not be added the burden of hardware maintenance, software installation, configuration, operation, or "pulling wire." It is simply a poor use of highly trained personnel. It also distracts them from the important work of improving the teaching-learning process.

Third, although it is important administrators maintain their information and knowledge base on the numbers of actual computer equipment for forecasting, staffing, insurance, and budgeting purposes and increase their knowledge about the types of equipment to best implement curriculum in schools and districts, administrators should not become the local "repair gurus" of our schools. Knowledge about what to "do" with technology is not necessarily knowledge "about" technology.

Fourth, even in states where "parallel"[1] certification has been enacted, it is not prudent to label individuals "administrators" if they really see themselves as technicians. To make it a requirement that highly qualified individuals pass through a program for administrator certification just to fix a school's computers or networks is simply a waste of valuable educational resources—for both K-12 settings and higher education. It is a way to increase higher education enrollments at a time when resources for maintaining huge cadres of faculty and graduate students are diminishing. This additional educational burden may screen out some highly qualified individuals from providing the necessary services (on a full-time basis) that schools need. It also places on these individuals the burdens of those who really wish to be administrators—the responsibilities historically known as "buildings, busses, and budgets." Oh, yes, there are arguments that there is a cost benefit in contracting out (or "outsourcing") such work, but there have been no substantial studies to provide definitive data on this proposition.

The reality is that schools do not keep large inventories of equipment to fill in for machines that are out for repair or experiencing "downtime." Educational settings, especially K-12, do not possess the resources to provide backup systems to compensate for those brought off-line for routine maintenance or software upgrades—especially necessary for on-line servers and routers that run 24 hours per day. Most schools obtain only sufficient levels of equipment to operate, and these are garnered from single-time grants, state-driven buy/purchase programs, discards from large local universities, or onetime school-business partnerships (Glennan & Melmed, 1996). When school machines do not function, teachers and students do without and revert to time-tested methodologies—paper and pencils, blackboards, workbooks, and so on. Although there is surface logic to this approach, on closer inspection we notice the system does not provide children the skills employers already say employees must have—the ability to communicate via electronic means for the conduct of business; the ability to clearly communicate in writing via word processing; the ability to teleconference to reduce costs for travel; the ability to conduct business worldwide across geographical and political boundaries without leaving the local business site; and the ability to troubleshoot with field counterparts and simultaneously view problems and identify and develop alternative solutions on-line as well as distribute corporate information, personnel policies, and data in real-time modes. None of this is possible when schools have off-line, malfunctioning, or nonexistent technologies (Glennan & Melmed, 1996). Without the capacity for repair or renewal, new technologies simply die as too unwieldy and take the road to demise as did instructional television earlier in this century (Cuban, 1986).

The *Washington Times* excerpt would be sad were it not for he fact that it is not unique. Some see the latest move to incorporate technology as a dangerous panacea for the "Let's improve scores in our schools" movement:

All the hoopla around the Internet obscures the deeper and more important issues of learning about how do you teach kids to acquire the basic skills and to think independently. . . . It's what I call the romance with the machine, and it has happened before. . . . It's driven by this dream of a magical solution that does not exist. (Cuban, cited in "Net Day Volunteers . . . ," 1997)

Although it may be dangerous to attribute traits and abilities to technology that it simply cannot be held to, it is just as dangerous to dismiss it as a useless tool destined for the educational junk pile of failed innovations. As with our national infrastructure, the past decade has seen the reduction of school district capital budgets for maintenance seriously erode the stability of the educational infrastructure (U.S. Congress, Office of Technology Assessment, 1995). In addition, during this same time period, communities, via local boards of education and state legislatures, have pressed for increased diligence in overhead, staffing, and personnel cost reductions so greater percentages of available dollars can be redirected toward instructional purposes. Unfortunately, this "reductionist" mentality coincides with market demands for greater technology access in our nation's schools (Glennan & Melmed, 1996). During this same time frame, the press for increased technology has visited our institutions of higher education. As we see in the chapters by Bromley and Jacobson; Waugh and

Handler; and LaCost, Seagren, and Stick (this volume), the resources required of institutions of higher education for the training of administrators and pre-service teachers carry differing administrative weights, with mixed sets of institutional messages, standards, and priorities across institutions, and face resistance from faculty for in-service, curricular integration, and migration issues, as well as traditional resource distribution issues.

In 1995, the U.S. Bureau of Labor Statistics (1996) surveyed employers who provided training. The resulting report documented that approximately 96% of employees received some type of informal training from employers. Computer training accounted for 38% of formal and 54% of the informal training received. Furthermore, between the months of May and October 1995, computer training accounted for more hours than both formal and informal training—nearly 11.8 hours, of which 5.1 hours were in formalized instructional classes.

During this same time, the U.S. deficit for goods and services increased to $10.4 billion (August 1997) from $10 billion (revised/July) due to continuing increases in imports over exports. The largest increase appeared in capital goods—primarily computer accessories, telecommunications equipment, and semiconductors as well as industrial supplies and materials. The technology goods and services trend is unmistakable. Our workforce is unable to compete on the world market. The country that invented the computer chip is now importing more high technology than it exports—and we are not training our future citizens to even use it, let alone to reverse the trend. While we in the United States debate if technology should be an integral part of our educational system, the world market demand for technology-related skills, goods, and services drives the national imbalance of trade to ever higher levels (U.S. Congress and U.S. Census Bureau, 1997). Rather, the debate must turn to how to fully exploit available systemic resources.

The Economics of Ignorance

There are at least two levels on which our system of resource ignorance operates.[2] First is at the local level. Here, we have the issues raised by the authors of several chapters in this volume (Crampton,

Nelson, Angevine), as well as by others (Glennan & Melmed, 1996; National Society for the Study of Education, 1996; Westbrook & Kerr, 1996). Second are those that take an ever-increasing center on the global education stage as documented by McClure, LaCost et al., Dizdar and Wandiga, and Vargas-Baron (this volume). The economics of ignorance refers to unintended consequences resulting when institutions, locales, or political entities bypass or overlook the global communities of education and how such communities through electronic linkages can reshape the nature of education—and the nature of global relationships. Westbrook and Kerr (1996) looked at the various plans and patterns that technology funding has taken and proposed a system that could produce a more "inclusive" system so that the reallocation of existing resources might be more aligned with the existing educational and political priorities of a state. They echoed the propositions and conclusions drawn by Glennan and Melmed (1996) that today's schools have not been, nor are likely to be, on the cutting edge of technologic breakthroughs. But the historic reliance on local largess, grants, or minimal reallocation of local resources does not have to be the deciding factor in the expansion of technology. Glennan and Melmed (1996) summarized the singular most limiting factor in the K-12 environment as:

> symptomatic of a deeply ingrained problem of social service providers in general and educational agencies in particular. Compared with the private sector, they lack an investment mentality. School districts do not regularly set aside a specified portion of their revenues for investing in activities to improve school performance. The reasons for this are found in the political nature of resource allocation in public education. (p. 2)

Here the economics of ignorance take the form of, "What is overwhelming cannot be overcome"—or possibly, "No investment is a sound investment." Local school communities and boards of education, overwhelmed by the rapid transitions in hardware and software development, cannot visualize how to meet rapidly changing needs. The private sector, with its experience and expertise, must step up to the plate—and not just for "onetime" donations of outdated inventory write-offs. The investment and wise allocation of funds for the provision of technology need to be addressed. In their chapter, Waugh and Handler (this volume) discuss the success of one school district's use of the Rule of Thirds. There is no reason a Rule of Thirds, Fourths, or Fifths could not be incorporated into the normal budget cycle of all schools. Additionally, schools are hampered by a highly politicized and vulnerable decision and budgetary process, especially in areas of low expertise, such as technology. Once again, Glennan and Melmed (1996) point out that to create technology-rich schools and avoid the dangers endemic to false economies

> causal links between investment actions and learning outcomes are lacking, and the latter are difficult to measure. . . . Private organizations might be able to use expert judgments in evaluating the alternatives. . . . (p. 2)
> [If we] assume that such schools might have a continuing annualized cost of $450 per student for hardware, additional personnel, software and materials, and training, this cost constitutes nearly 8 percent of the expected national average, current, per-student expenditure of $5,600—or about 12 percent of the resources allocated directly to the school building. Clearly, this poses a different magnitude of difficulty. (p. 6)

Recently, ATTCapital released results of a commissioned study of more than 80,000 U.S. school districts and found that 97% of districts nationwide now have a technology plan in place, but 4 out of 10 indicated a lack of financial assistance to implement these plans. These disparities arose from lower K-12 operating budgets and reduced levels of federal and state support (ATTCapital, n.d. a). In light of Glennan and Melmed's (1996) pronouncement that schools should spend approximately $450 per student per year, the ATTCapital study suggests a figure of approximately $335 per student per year for at least the next 3 years to keep schools current with changing technology. Most alarming of all was a finding that 59% of the districts reported older computers confined to laboratories or administrative uses (ATTCapital, n.d. a). The study was commissioned by the CEO Forum on Education and Technology, based in Washington, D.C., as part of a 4-year effort to monitor technology's impact on student achievement. Table 11.1 presents the results for a district of between 500 and 5,000 students.

TABLE 11.1 Results of the CEO Forum on Education and Technology Study

No. of Students	@$335/student/year	@$450/student/year
500	$167,500	$225,000
1,000	$335,000	$450,000
2,500	$837,500	$1.12M
5,000	$1.67M	$2.25M

Clearly then, this approach is inextricably entwined with the fiscal resources need. It can lead to a lack of a technology-driven focus or may eliminate technology from consideration completely. Glennan and Melmed (1996), reporting on the supply and demand of educational software, found that the educational software market totaled approximately $750 million—about 0.3% of all K-12 expenditures. They also noted that software developers believed the educational K-12 sector did not place highly sophisticated demands on software, purchased fewer site license/units, and exhibited an inability to generate economies of scale in the use of content-rich software unlike industrial counterparts. Even at an estimate of $50 per copy, the educational market would only generate $5 million if all the 100,000 U.S. schools purchased a single copy, whereas if that same CD-ROM sold to only 1% of all U.S. households it would generate revenues of $50 million for its copyright holders and distributors (Glennan & Melmed, 1996). The economies of scale simply are not there.

Whither Thou Goest . . .

The second proposition is at the international, or global, level of concern. Dizdar and Wandiga (this volume) document how technology is helping to revitalize the almost total educational devastation visited upon one country. Vargas-Baron takes us on another route, one that builds on economic relationships through our institutions of higher education and their alumni around the world and takes advantage of their roles at home in ministries by developing trade and exchange agreements. Their economic arguments for utilizing technologic means to accomplish educational ends has great power—

especially when coupled with the stories "from the battlefield" that accentuate the needs experienced by the children and teaching professionals of countries around the globe. The power provided to K-12 schools, community colleges, and institutions of higher education when they band together has immense potential for change both in the United States and around the globe. This globalization and public-private partnering could substantially change the directions of schooling, as well as markets, in the coming decades. It will take the continued efforts of the authors of this volume and many others to bring these many dreams to reality. It requires a deeper commitment from the private sector, as well as from local, state, and national governments. Although some argue that technology is a panacea diverting us away from the real issues of school improvement and reformation, the authors in this volume see matters quite another way. Without increased use of technology as a "tool" for the improvement of the economic situation of all children throughout the world, there will be no continued improvements. For without such a commitment, schools will produce graduates unprepared for the demands of the global political and economic marketplace. They will be without the intellectual, workforce, and diplomatic means for preserving and enriching their own lives or the lives of their heirs. Without technological improvements—without change and growth—global communities will wither before they bloom. The authors here have provided us with strong, vibrant road maps to explore the information superhighways. Let us echo Franklin Delano Roosevelt when he said that there is no fear worse than fear itself. The authors of this AEFA yearbook have shown us a clear and vital path in the future of technology. Let us not allow fear in any of its forms (i.e., fear of failing, fear of change, fear of technology, etc.) to entrap and prevent our implementation of this unique vision. Let us begin now to help our schools, our faculties, and our global markets by beginning one small step at a time.

Notes

1. Parallel certification is granted by states to individuals whose work experience or professional credentials have given them "equivalent" experience to the training and duties of school personnel while working in the private sector. Upon application and

review by state certification officials, an individual may be granted a permanent, or a temporary, certificate to work as a teacher or administrator in a K-12 school environment by the state.

2. *Resource ignorance* here is intended to mean the overlooking or ignorance of the computer technology resource allocation issue, i.e., a lack of attention to, not the lack of knowledge about the issue. Of course, in selected circumstances a lack of actual knowledge may add to an individual's or organization's neglect due to a lack of complete understanding.

References

ATTCapital. (n.d. a). http://www.attcapital.com/corporate/art_schools.htm.

ATTCapital. (n.d. b). http://www.attcapital.com/corporate/technology.htm.

Cuban, L. (1986). *Teachers and machines: The classroom use of technology since 1920.* New York: Teachers College Press.

Glennan, T. K., & Melmed, A. (1996). *Fostering the use of educational technology: Elements of a national strategy.* RAND Corporation (http://www.rand.org/publications/MR/MR682/contents.html).

National Society for the Study of Education. (1996). *Technology and the future of schooling* (95th yearbook). Chicago: University of Chicago Press.

Net Day volunteers seek to plug more schools into the web. (1997, October 25). *Chicago Tribune.* New York Times News Service.

U. S. Congress, Office of Technology Assessment. (1995, April). *Teachers and technology: Making the connection* (OTA-EHR-616). Washington, DC: Government Printing Office.

U.S. Congress & U.S. Bureau of Labor Statistics. (1996, December). *1995 survey of employer provided training—employee results. Washington, DC: Government Printing Office. (http://stats.bls.gov/news.release/sept.nws.htm)*

U.S. Congress & U.S. Census Bureau. (1997, October). *U.S. trade in goods and services highlights.* Washington, DC: Government Printing Office. (http://www.census.gov/indicators/www/ustrade.html)

Washington Times. (1997, June 24). [On-line]. http://www.washtimes.com/

Westbrook, K. C., & Kerr, S. T. (1996). Funding educational technology: Patterns, plans, and models. In National Society for the Study of Education, *Technology and the future of schooling* (95th yearbook, chap. 3). Chicago: University of Chicago Press.

American Education Finance Association Board of Directors, 1997–1998

OFFICERS

Eugene P. McLoone, *President*
R. Craig Wood, *President-Elect*
Lawrence O. Picus, *Immediate Past President*
George R. Babigian, *Executive Director*

DIRECTORS

1998 Term

Patrick F. Galvin
Carolyn D. Herrington
Maureen McClure
John Schneider
Stephanie E. Stullich

1999 Term

Mary F. Fulton
Robert K. Goertz
Stephen L. Jacobson
Richard A. King
Barbara LaCost

2000 Term

Jonathan Lambertson
John McEwen
Jennifer King Rice
Catherine C. Sielke
Leanna Stiefel

Sustaining Members

Jewell C. Gould
Edward J. Hurley
Chris Malkiewich
Michael Resnick
Donald I. Tharpe

Mary F. Hughes, *Editor,*
Journal of Education Finance

Index

CORWIN
PRESS

The Corwin Press logo—a raven striding across an open book—represents the happy union of courage and learning. We are a professional-level publisher of books and journals for K–12 educators, and we are committed to creating and providing resources that embody these qualities. Corwin's motto is "Success for All Learners."